Transactions on Computer Systems and Networks

Transactions on Computer Systems and Networks is a unique series that aims to capture advances in evolution of computer hardware and software systems and progress in computer networks. Computing Systems in present world span from miniature IoT nodes and embedded computing systems to large-scale cloud infrastructures, which necessitates developing systems architecture, storage infrastructure and process management to work at various scales. Present day networking technologies provide pervasive global coverage on a scale and enable multitude of transformative technologies. The new landscape of computing comprises of self-aware autonomous systems, which are built upon a software-hardware collaborative framework. These systems are designed to execute critical and non-critical tasks involving a variety of processing resources like multi-core CPUs, reconfigurable hardware, GPUs and TPUs which are managed through virtualisation, real-time process management and fault-tolerance. While AI, Machine Learning and Deep Learning tasks are predominantly increasing in the application space the computing system research aim towards efficient means of data processing, memory management, real-time task scheduling, scalable, secured and energy aware computing. The paradigm of computer networks also extends it support to this evolving application scenario through various advanced protocols, architectures and services. This series aims to present leading works on advances in theory, design, behaviour and applications in computing systems and networks. The Series accepts research monographs, introductory and advanced textbooks, professional books, reference works, and select conference proceedings.

Ashok Kumar Yadav · Ali Ahmadian · Ajay Pratap

Multi-attribute Decision-Making: Bridging Theory and Practice

Springer

Ashok Kumar Yadav
Department of Information Technology
Rajkiya Engineering College Azamgarh
Deogaon, Uttar Pradesh, India

Ajay Pratap
Department of Computer Science
and Engineering
Indian Institute of Technology Varanasi
Varanasi, Uttar Pradesh, India

Ali Ahmadian
Faculty of Engineering and Natural
Sciences
Istanbul Okan University
Istanbul, Turkey

ISSN 2730-7484 ISSN 2730-7492 (electronic)
Transactions on Computer Systems and Networks
ISBN 978-981-95-4781-4 ISBN 978-981-95-4782-1 (eBook)
https://doi.org/10.1007/978-981-95-4782-1

This Springer imprint is published by the registered company Springer Nature Singapore Pte Ltd.
The registered company address is: 152 Beach Road, #21-01/04 Gateway East, Singapore 189721, Singapore

Ali Ahmadian would like to dedicate this work to his lovely wife, Ferial for her endless support.

Foreword

It is with considerable enthusiasm that I introduce *Multi-attribute Decision-Making: Bridging Theory and Practice.* In an age marked by unprecedented complexity and interconnectedness, the ability to make sound, reasoned decisions is not merely an advantage but a necessity. The challenges we face, whether in technological advancement, resource management, policy development, or strategic planning, invariably involve navigating a landscape of multiple, often competing, objectives and constraints. The field of Multi-attribute Decision-Making (MADM) provides the essential scientific framework and methodological tools to address such challenges systematically and effectively.

This forthcoming manuscript stands as a commendable and timely contribution to the MADM literature. The authors have undertaken the ambitious task of creating a work that is both comprehensive in its scope and accessible in its presentation, catering to a diverse audience ranging from students embarking on their journey into decision science to seasoned practitioners seeking to refine their analytical toolkit. What particularly distinguishes this volume is its holistic approach to the decision-making process. It meticulously guides the reader through each critical stage, from the foundational principles of decision theory and the often-underestimated significance of data normalization to the sophisticated intricacies of criteria weighting and the application of various ranking models.

The emphasis placed on criteria weighting is especially noteworthy. The process of assigning importance to different criteria is a linchpin of any MADM application, and this book offers an admirably thorough treatment of the subject. By dedicating distinct chapters to subjective methods that leverage expert intuition, objective methods that derive insights from data patterns, and integrated approaches that seek a synergistic balance, the authors provide a nuanced understanding of this critical phase. This comprehensive perspective is invaluable for fostering robust and defensible decision outcomes.

Moreover, the commitment to "Bridging Theory and Practice" is woven throughout the manuscript. This is not just a theoretical treatise; it is a practical guide designed to empower readers to apply these powerful analytical techniques to real-world problems. The planned inclusion of chapters detailing diverse applications and exploring future research avenues further solidifies this commitment, ensuring that the knowledge imparted is both relevant and forward-looking.

Multi-attribute Decision-Making: Bridging Theory and Practice is poised to become a seminal text in the field. It offers a lucid, well-structured, and insightful exploration of MADM, equipping its readers with the understanding and tools necessary to confront complex decisions with enhanced clarity, rigor, and confidence. I congratulate the authors on this significant achievement and am confident that this work will be highly valued by the academic and professional communities alike, inspiring both sound application and further innovation in this vital domain.

Reggio Calabria, Italy Massimiliano Ferrara, Ph.D.

Preface

The landscape of decision-making in our contemporary world is characterized by an ever-increasing complexity, a multitude of stakeholders, and an abundance of information, often fraught with uncertainty. In such an environment, the ability to make rational, justifiable, and effective choices is paramount. Multi-attribute Decision-Making (MADM) offers a robust and structured paradigm to navigate these complexities, providing a suite of methodologies to evaluate alternatives against numerous, often conflicting, criteria. This manuscript, *Multi-attribute Decision-Making: Bridging Theory and Practice*, has been developed out of a perceived need for a comprehensive guide that not only elucidates the theoretical underpinnings of MADM but also demonstrates its practical application across a wide spectrum of real-world challenges.

Our primary motivation for this work is to provide a cohesive and accessible resource for students, academics, researchers, practitioners, and policymakers. We aim to demystify the often-intricate processes involved in MADM, from the initial structuring of a decision problem to the final selection and validation of an optimal alternative. A significant emphasis is placed on the foundational elements that are critical for sound decision analysis. This includes a thorough exploration of normalization strategies, essential for rendering diverse data comparable, and an in-depth examination of criteria weighting—a pivotal step that profoundly influences outcomes. We dedicate substantial portions of this book to dissecting Subjective Weight Methods, which harness the invaluable insights of expert judgment; Objective Weight Methods, which derive criteria importance from the inherent structure of the data; and Integrated Weight Methods, which thoughtfully combine both perspectives to achieve a balanced and robust assessment.

The journey through this book is designed to be progressive. We begin in Chap. 1 by laying the groundwork with an introduction to decision theory and the core concepts of MADM. Chapter 2 delves into the critical role and various techniques of normalization. Chapters 3–5 are dedicated to a comprehensive treatment of subjective, objective, and integrated weighting methodologies, respectively. Following this, Chap. 6 explores a variety of MADM ranking models, comparing their operational mechanics, strengths, and limitations. Chapter 7 will showcase the versatility of

MADM through a discussion of its applications in diverse fields such as sustainable development, smart city planning, healthcare management, energy planning, disaster management, and advanced telecommunications. Finally, Chap. 8 will address the current research challenges and exciting future directions in the field, aiming to stimulate further inquiry and innovation.

This manuscript is the culmination of extensive research, practical experience, and valuable interactions with colleagues and students. We are indebted to the pioneers and contemporary researchers in the field whose work has laid the foundation upon which this book is built. It is our sincere hope that *Multi-attribute Decision-Making: Bridging Theory and Practice* will serve as an enlightening and practical guide, empowering readers to confidently navigate the complexities of modern decision-making and to effectively translate sound theory into impactful practice. We believe that a thorough understanding and judicious application of MADM techniques can lead to more informed, equitable, and sustainable outcomes across all domains of human endeavor.

Deogaon, India — Ashok Kumar Yadav
Istanbul, Turkey — Ali Ahmadian
Varanasi, India — Ajay Pratap

Competing Interests The authors have no competing interests to declare that are relevant to the content of this manuscript.

Introduction

The act of decision-making is fundamental to human endeavor, permeating every aspect of personal life and professional responsibility. In an increasingly interconnected and complex world, individuals and organizations are continually faced with choices that have significant and often far-reaching consequences. Decision theory provides a structured and rational framework for navigating these choices, particularly when outcomes are uncertain. It encompasses a range of mathematical models, principles, and methods designed to analyze decision problems, evaluate potential outcomes, and guide the selection of the most desirable course of action based on defined preferences, available information, and the likelihood of various consequences. The essence of decision theory is not to foresee the future with absolute certainty, but rather to equip the decision-maker with a systematic approach to make the best possible choice given the information at hand, thereby minimizing risks and maximizing the potential for favorable results.

Within decision theory, problems can be broadly categorized based on the level of knowledge regarding their outcomes: decision-making under certainty (where outcomes are known), under risk (where outcomes are probabilistic and probabilities are known), and under uncertainty (where outcomes or their probabilities are unknown). Many elementary decision problems can be simplified by focusing on a single attribute or criterion, such as minimizing cost or maximizing profit. Such Single-Attribute Decision Models offer clarity and computational ease for straightforward situations. However, the vast majority of real-world decision problems are inherently multifaceted, involving numerous, often conflicting, objectives that cannot be adequately addressed by optimizing only one dimension. For instance, selecting a new technology might involve balancing performance, cost, user-friendliness, and environmental impact. This is where the limitations of single-attribute models become apparent, necessitating more sophisticated approaches.

Multi-attribute Decision-Making (MADM) emerges as a critical sub-discipline of decision theory specifically developed to address these complex scenarios. MADM provides methodologies to evaluate, rank, compare, or select alternatives when they are characterized by multiple, often incommensurable and conflicting, attributes or criteria. The genesis of MADM can be traced to the mid-20th century, with foundational work in operations research and economics, including concepts like Pareto optimality and utility theory, laying the groundwork. The development of pioneering methods such as the Analytic Hierarchy Process (AHP) and ELECTRE further solidified MADM as a distinct and powerful field of study.

The need for MADM is underscored by the inadequacy of traditional single-criterion optimization in real-world contexts. Decisions frequently require balancing diverse objectives, considering the varied preferences of multiple stakeholders, and navigating uncertainties. MADM offers a structured, transparent, and often quantitative framework to incorporate these complexities, allowing for a more holistic and justifiable decision-making process.

The general MADM process, which forms the backbone of this book, involves several crucial phases:

1. **Problem Definition and Structuring**: Clearly identifying the decision goal, the alternatives to be considered, and the relevant criteria for evaluation.
2. **Data Collection and Decision Matrix Construction**: Gathering the performance data of each alternative with respect to each criterion.
3. **Normalization of the Decision Matrix**: Transforming the data into a common, dimensionless scale to ensure comparability across criteria that may have different units or ranges. This vital step will be thoroughly explored in Chap. 2.
4. **Weighting of Criteria**: Determining the relative importance or significance of each criterion. This is a pivotal phase, as the assigned weights can profoundly influence the final ranking. Chapters 3–5 will delve into subjective, objective, and integrated weighting methods, respectively.
5. **Aggregation and Ranking**: Applying a specific MADM model to aggregate the weighted, normalized performance scores and derive a final ranking or selection of alternatives. A variety of these models will be detailed in Chap. 6.
6. **Sensitivity Analysis and Final Decision**: Assessing the robustness of the results to changes in inputs and making the final choice.

This book aims to provide a comprehensive journey through the theory and practice of Multi-attribute Decision-Making. Starting with this foundational introduction, we will systematically explore each of the key components and methodologies. Chapter 2 focuses on normalization techniques. Chapters 3–5 provide an in-depth examination of subjective, objective, and integrated (or comprehensive) weighting methods. Chapter 6 presents various MADM ranking models. Chapter 7 delves into the challenges inherent in MADM models and discusses future trends. Finally, Chap. 8 will illustrate the wide-ranging applicability of these techniques across diverse fields. Through this structured exploration, we endeavor to equip readers with the knowledge and tools necessary to effectively apply MADM principles to complex decision problems, thereby bridging the gap between robust theory and impactful practice.

Contents

Chapter 1
Introduction to Decision Theory

Abstract Decision theory encompasses mathematical models, principles, and methods that aids individuals, organizations, and policymakers to analyze decisions, evaluate potential outcomes, and select the most desirable course of action based on preferences, probabilities, and consequences. This chapter covers the fundamentals of decision theory, including single-attribute and multi-attribute decision models, an overview of fuzzy theory, the role of criteria and attributes in decision-making, and key weighting and ranking methods used in evaluations.

Keywords Decision · Single-attribute decision-making · Multi-attribute decision-making · Decision-making · Ranking · Fuzzy · Decision criteria · Decision matrix · Multi-criteria decision-making

1.1 Basics of Decision Theory

Making decisions in an environment of uncertainty is an ongoing part of human life. Whether in our personal lives or professional responsibilities, we frequently act without fully understanding the consequences of our decisions. This challenge is especially severe for systems engineers and managers, who must make critical decisions in complex, constantly shifting technological environments. In such cases, uncertainty is not just a factor; it is frequently the primary challenge. Human nature tends to avoid uncertainty. Scientific efforts are now focused on forecasting future trends, particularly in economics and technology. If predictions were assumed to be perfectly accurate, many decision problems would be reduced to optimization tasks. The optimal course of action could then be determined using techniques such as mathematical programming. However, this idealization ignores the critical element of uncertainty, which can lead to fundamentally flawed decisions. A decision only exists if there are at least two distinct alternatives that lead to different consequences. After making a choice and realizing its outcome, reversing it typically results in costs or penalties. A robust decision-making theory should provide a structured way to use all available information and identify the alternative that best aligns with

A. K. Yadav et al., *Multi-attribute Decision-Making: Bridging Theory and Practice*, Transactions on Computer Systems and Networks,
https://doi.org/10.1007/978-981-95-4782-1_1

the decision-maker's preferences. However, it is essential to distinguish between a good decision and a good outcome. A well-reasoned decision can still lead to poor results due to factors beyond our control, just as a poorly reasoned decision may occasionally succeed by chance. The decision theory is not about predicting the future with certainty. Rather, it is a systematic approach to making the best possible choices provided the information at hand. It minimizes the risk of poor outcomes but cannot eliminate all uncertainty. Decision theory formalizes common sense using mathematics, particularly through utility theory to represent preferences and probability theory to represent uncertainty [1].

Decision theory is a framework for making choices when facing uncertainty, considering various factors like available alternative, potential outcomes, and the probability of those outcomes. It helps individuals and organizations make the most rational and effective decisions by evaluating the potential benefits and risks associated with each alternative. Decision theory provides a rational framework for choosing between alternative courses of action when the consequences resulting from this choice are imperfectly known. Two streams of thought serve as the foundations: utility theory and the inductive use of probability theory. Decision theory is a theoretical framework for making rational choices under certainty, risk, or uncertainty that helps identify the best course of action based on preferences, available information, and possible outcomes. The decision theory can be of normative, descriptive, and prescriptive decision Theory which can focuses on how decisions should be made, analyzes how decisions are actually made and provides tools and techniques to improve decision-making, respectively Multi-attribute decision making (MADM) is a branch of decision-making theory that deals specifically with problems involving multiple, conflicting, and complex decision criteria with goal of rank, compare, or select alternatives based these conflicting decision criteria [2].

The decision theory is classified as shown in Fig. 1.1 based on the level of knowledge about outcomes and consequences into three main categories: decision-making under certainty, risk, and uncertainty. In the case of making decisions with certainty, all outcomes are known and deterministic. In such cases, decision methods are directly applied with complete information. For example, selecting a supplier based on known cost, quality, and delivery time. Decision-making under risk includes probabilistic outcomes, but the probabilities are known. In this case, decision-makers use expected utilities, probabilistic MADM approaches, or stochastic models. For example, choosing an investment portfolio with known risk-return profiles. Decision-making under uncertainty includes outcomes and their probabilities that are unknown. Methods like Maximin, Maximax, Minimax Regret, and fuzzy MADM are used. For example, selecting a technology under evolving standards and unpredictable regulations.

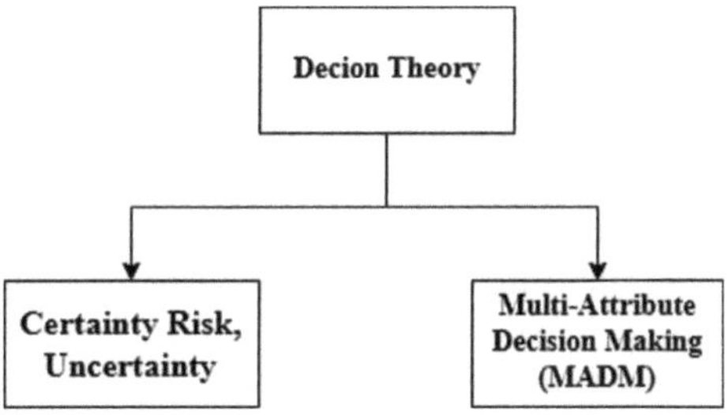

Fig. 1.1 Decision theory classification

1.2 Single-Attribute Decision Models

Decision-making is a fundamental process that involves multiple actors, including individuals, groups, institutions, and governments. Multi-criteria decision-making, as an academic discipline, is relatively young but has recently gained significant importance in addressing complex decision problems in the fields of science, engineering, and economics by evaluating multiple conflicting criteria to support informed and balanced choices across diverse domains. Since the 1950s and 1960s, when the foundations of modern multi-criteria decision-making methods were established, numerous researchers have dedicated their efforts to developing new models and techniques to enhance decision-making processes across various fields. Over the past few decades, research and development in this field has advanced rapidly and exhibits no signs of slowing, with progress appearing to be increasing at an exponential rate. This rapid advancement is due to the applicability of multi-criteria decision-making methods in the fields of science, engineering, economics, and policymaking. Despite significant worldwide advancement and intensive development, there have been limited systematic efforts to compile and present the theoretical foundations and advances in multi-criteria decision-making methods. However, the methodological choices and framework for assessing decisions are still under discussion [3].

The single-attribute decision models generally have four steps to find the optimal solution namely: define the attribute, gather data, make a comparison, and select best alternative. The first step identifies the criterion that will influence the decision. The second phase gather data used to collect information about the chosen decision criteria for all available alternative. The third phase make a comparison of all alternative based on the chosen decision criteria. And the final step selects the best alternative with respect to the defined decision criteria. Single-attribute decision models are decision-making frameworks where decisions are based on a single decision criterion. These models are useful when a decision must be made quickly, or the complexity needs to be minimized, or the decision is straightforward, with one attribute being the most significant for the outcome. This model revolves around a single key factor, ignoring other attributes. For example, in purchasing decisions, one might focus solely on price, ignoring factors like quality or brand reputation. These models are particularly useful in situations where the decision-maker is interested in optimizing a single factor such as cost, time, profit, or efficiency without considering other attributes.

While real-world decisions often involve multiple criteria, simplifying the process to one attribute can aid clarity, speed, and ease of computation. Single-attribute decision models are best suitable models when an attribute in the decision problem overwhelmingly influences the decision or other attributes are either constant or insignificant or time or resource constraints limit the scope of evaluation.

While single-attribute decision models offer a straightforward and practical approach to decision-making. They are best suited for problems where one attribute clearly dominates. In complex decision problem scenarios, decision-makers should be cautious about relying solely on these models and consider multi-attribute frameworks when necessary. Nonetheless, they serve as valuable tools for initial assessments, simplifying complex choices into manageable decisions. On other hand single-attribute decision models may ignore important aspect in taking decision which may lead in suboptimal decisions. This approach is based on over-reliance on a single decision criterion that may reflect personal or organizational biasness. This biasness may affect real-world decisions which often involve changing variables, which single-attribute models may not account for. These approaches have the high probability of poor decisions if the single decision criteria are not comprehensive and it can introduce subjectivity. In other word we can say that single-attribute decision models may not reflect true preferences have limited applicability along with possibility of Biasness and lack of trade-offs. These are some of the crucial challenges inherent in the single-attribute decision models that make them incapable of addressing complex decision problems that have dependency on more complex and conflicting decision criteria. To address the challenges of single-attribute decisions, multiattribute decision-making (MADM) models have been proposed. In the next section, we discuss in detail the multi-attribute decision-making models. The scope of the book is mainly focused on the discussion on this MADM.

1.3 Multi-Attribute Decision-Making

The art of multi-attribute decision-making approach is closely linked to the principles of rational choice theory, as both focus on selecting the best alternative based on logical evaluation. It assumes that individuals are driven by financial gain and profit potential, enabling the development of formal, and often predictive, models of human behavior. People generally act rationally within certain constraints and according to the information available to them about their circumstances. However, human actions also include elements that are not purely rational. Rational choice theory holds that individuals evaluate possible potential alternatives, anticipate their outcomes, and select the best alternative they believe will yield the most favorable result for themselves. Since people cannot have everything at once, they must prioritize and decide which goals to pursue and how to achieve them. This process requires balancing available resources, time, and effort to select the most realistic and rewarding options while considering the limitations and trade-offs involved. Rational individuals select the option that they believe will provide them with the highest level of satisfaction.

Although many researchers contributed to the development of the expected utility model, Von Neumann and Morgenstern (1947) are widely credited with establishing its first rigorous axiomatic basis. Their work laid the groundwork for studying how rational people make decisions in an environment of uncertainty, with profound consequences for economic theory, game theory, and decision sciences. Today, the expected utility model serves as a fundamental normative framework for decision analysis [3].

Over the past several decades, operations research has significantly evolved as a key discipline in supporting scientific management. It primarily focuses on developing models and employing algorithmic optimization techniques to analyze and solve complex real-world problems. Since the foundational works of von Neumann and Morgenstern (1947) and Savage (1954), this approach has emerged as the leading paradigm for decision analysis and support, particularly when multiple evaluation criteria are involved. The traditional operational techniques serve the same purpose, which is to maximize or minimize a utility function while considering constraints. Multi-criteria decision making (MCDM) can be considered both a conventional and a modern aspect of operations research, depending on one's perspective. It has emerged as one of the most rapidly expanding problem areas across various disciplines. The core problem is how to evaluate a set of alternatives in terms of a number of conflicting decision criteria. Many modern researchers have considered MCDM problems. MCDM refers to making decisions in the presence of multiple, usually conflicting, criteria. There is no unique and well-defined methodology that one could follow step-by-step from the beginning to the end of a decision aiding process. When working with objects that must be described and compared based on multiple characteristics, aggregation becomes a critical challenge. It involves synthesizing these often-conflicting decision criteria to achieve objectives such as selecting the best alternative, ranking the alternatives, or classifying them into distinct categories.

MCDM methods cover a wide range of quite distinct approaches. These methods are generally divided into two main categories: discrete Multi-Criteria or MADM and continuous multi-objective decision making (MODM). They are used to evaluate and determine which option or course of action is more desirable based on multiple criteria. Each problem has multiple, usually conflicting objectives or decision criteria. Each objective or decision criterion has a different unit of measurement. MCDM can be understood as the process of evaluating real-world scenarios using a combination of qualitative and quantitative decision criteria under conditions of certainty, uncertainty, and risk, with the goal of identifying the most suitable course of action, alternative, strategy, or policy from among multiple available alternative [3].

MADM is a core concept in decision theory used to evaluate, rank, and select from a set of alternatives that are characterized by multiple, conflicting, decision criteria. It is widely used in engineering, management, economics, environmental science, healthcare, where decisions involve complex trade-offs. It is a set of procedures used to compare and rank a finite number of decision alternatives when each alternative is evaluated against multiple decision criteria. The concept of MADM emerged in the mid-twentieth century as a response to the limitations of single-criterion optimization in complex decision problems. The initial work (1950s–1960s) in operations research and economics laid the foundation, including Pareto optimality (Vilfredo Pareto) and utility theory (John von Neumann and Oskar Morgenstern) which further formalized and developed as MADM frameworks such as Elimination Et Choix Traduisant la Réalité (ELECTRE) by Bernard Roy for outranking methods in 1968 and Analytic Hierarchy Process (AHP) by Thomas Saaty for hierarchical decision structuring in 1977. A general classification of MADM has been illustrated in Fig. 1.2 presents the key components of MADM. Each of these components will be discussed in detail in the following chapters to provide a comprehensive understanding of MADM.

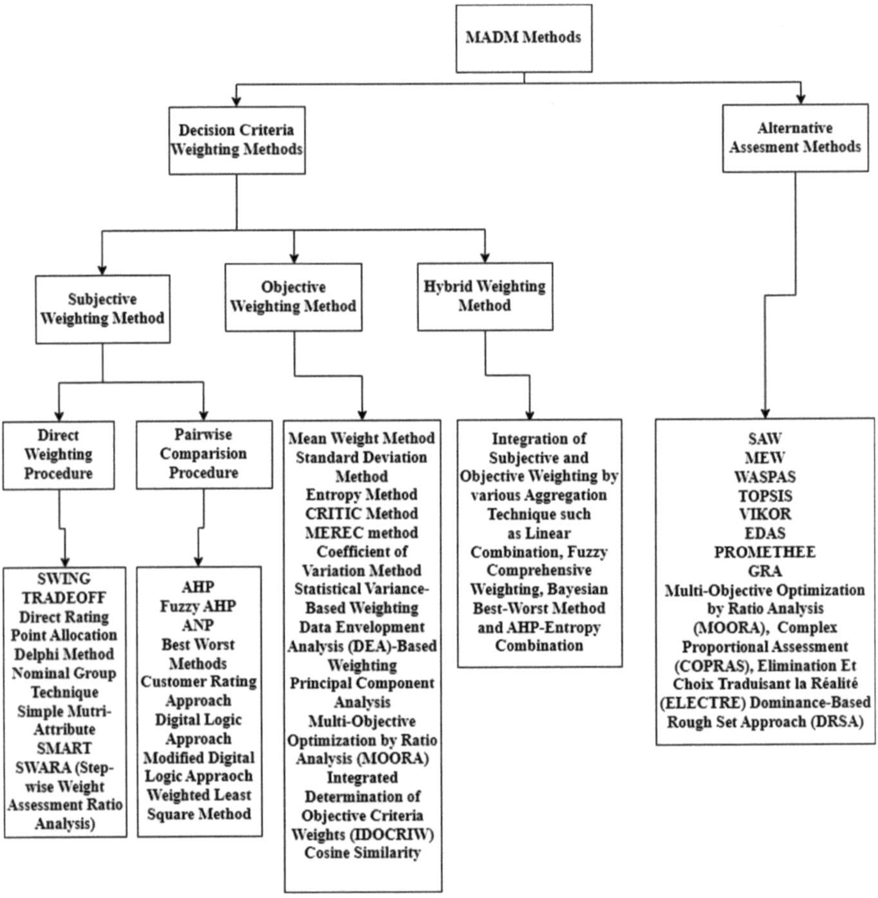

Fig. 1.2 General classification of MADM

1.4 Need of MADM

Traditional single-criterion decision-making fails in real-world scenarios where multiple conflicting objectives, such as benefit, cost, and quality, exist. These scenarios may involve uncertainty and subjectivity influencing decisions.

It also faces problems due to having different preferences of stakeholders, sustainability, and ethical considerations. Decisions require balancing conflicting objectives, with various stakeholders holding different priorities and preferences. This underscores the need for transparency and traceability in the decision-making process. MADM addresses these challenges through incorporating both quantitative evaluations and subjective judgments.

1.5 Component of MADM

The key components of MADM include alternatives, decision criteria, decision matrix and significance of each decision criteria, i.e., Weight. MADM have different options or choices such as different smartphones, for evaluation performance of each alternative have a set of decision criteria such as cost, screen size, battery backup. A decision matrix is a structured data that used to compare different alternatives based on multiple decision criteria. MADM need to assess the significance of each decision criteria for evaluating performance score of each alternative to rank, and select the best alternative. Table 1.1 presents a general component commonly involved in any MADM technique used for addressing complex decision problems. It provides a precise description of each component, highlighting its role and significance in the multi-attribute decision-making process.

Table 1.1 Components of MADM in decision making

S. no	Component	Description
1	Alternative	The different options or choices available
2	Decision criteria	The factors or dimensions used to evaluate the alternatives (e.g., cost, speed)
3	Decision matrix	A table showing how each alternative performs with respect to each attribute
4	Significance of decision criteria (weight)	Importance assigned to each criterion (based on user preference or analysis)

1.6 General Step in MADM

In general, the MADM technique include defining the decision problem, identifying decision-makers, objectives, and constraints, listing feasible alternatives, selecting evaluation criteria, constructing, and normalizing the decision matrix, assigning appropriate weights according to significance of each decision criteria, applying a ranking method, analyzing results, and making decisions. A generalized MADM technique involves the following Phases:

1.6.1 Define the Problem and Objectives

This step involves clearly outlining the decision problem and the main objective of the decision-making problem, including identification of decision-makers, objectives, and constraints influencing the decision-making process.

1.6.2 Identify Alternatives

Enumerate all viable alternatives that need to be assessed within the decision-making process, ensuring each option meets the defined criteria and constraints for a comprehensive and meaningful evaluation.

1.6.3 Determine Criteria

Identify and define the relevant decision criteria, both quantitative and qualitative, that will be used to evaluate each alternative, ensuring comprehensive and balanced assessment across all critical aspects of the problem.

1.6.4 Construct the Decision Matrix

In this step, ask them to develop a table where each row corresponds to a specific alternative and each column represents a decision criterion. This table is known as a decision matrix, which lists the performance of each alternative on each criterion. This performance matrix consists of how each alternative scores or performs with respect to every considered evaluation criterion identified in the above steps.

1.6.5 Normalize the Matrix

Normalization is one of the crucial steps in the MADM procedure at the time of computing the significance of each decision criterion and at the time of applying any ranking methods. The normalization of the decision matrix is required if scales differ in terms of decision criteria measured in different units and differ in their value range to convert values to a predefined common scale, usually [0, 1].

1.6.6 Assign Weights to Criteria

Assessment of the significance of each decision criterion in the selection of the best alternative is one of the most crucial and important aspects of the MADM procedure. This system ensures the consistency of the ranking. Determine importance weights to reflect the importance of each decision criterion. It is also necessary to test the robustness of the assigned weight of each decision criterion in the case of the subjective weighting method by varying the comparison matrix.

1.6.7 Apply a MADM Method

This step needs to select an appropriate decision-making method to systematically evaluate and rank considered alternatives based on defined decision criteria and the obtained significance of each decision criterion. This process helps identify the most suitable alternative among several choices by applying a structured approach that considers multiple factors, ensuring objectivity, consistency, and rational judgment in complex decision-making scenarios by computing the performance score of each alternative.

1.6.8 Rank the Alternatives

This is the final step, where the best option is chosen based on performance scores. These scores are calculated using a proper weighting and ranking method, along with a suitable normalization technique, to help identify the most optimal alternative. The above phases of the MADM techniques have been depicted in Fig. 1.3.

This book emphasizes the four essential phases of MADM: (1) constructing the decision matrix by defining alternatives and decision criteria, (2) normalizing the decision matrix, (3) determining the significance of each criterion through appropriate weighting, and (4) computing performance scores to rank and select the optimal alternative. Each of these phases is explored in depth in the following chapters.

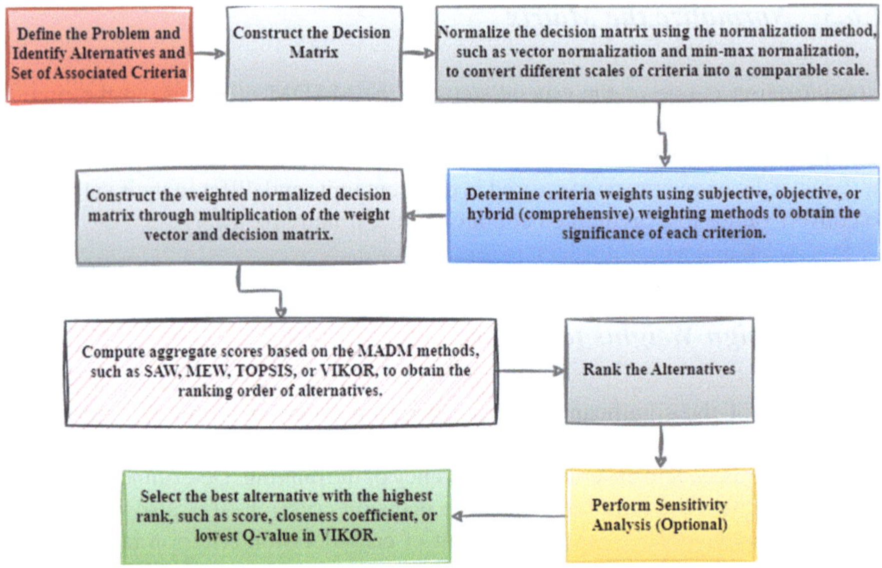

Fig. 1.3 General steps to compute ranking order and select best alternative in MADM

Chapter 2 focuses on normalization techniques and analyzes their influence on the consistency of ranking outcomes. Chapters 3, 4, and 5 provide a detailed examination of subjective, objective, and hybrid weighting methods, respectively, highlighting their theoretical foundations and practical implications. Chapter 6 presents a comprehensive discussion on various ranking methods used to evaluate and prioritize alternatives. Chapter 7 provides a comprehensive overview of the diverse applications of MADM techniques across various domains, illustrating their versatility and practical relevance in real-world decision-making scenarios. The final chapter, Chap. 8, explores the current research challenges associated with MADM methods, along with emerging trends and potential future directions in the field. This chapter aims to inspire further research and innovation by highlighting gaps in existing methodologies and discussing opportunities for advancing MADM frameworks to address complex decision problems in evolving technological and societal contexts. Through systematically addressing each phase and application, challenges, current trends, and future scope of MADM techniques this book provides a complete understanding of MADM methodologies, guiding readers in applying them effectively to real-world decision-making scenarios.

1.7 Basics Overview of Fuzzy Theory

Fuzzy theory, introduced by Lotfi A. Zadeh in 1965, provides a mathematical framework for handling uncertainty, vagueness, and imprecision, which are often present in real-world decision-making scenarios. In the context of MADM, fuzzy theory plays a crucial role when decision-makers face ambiguity in criteria weights, alternatives, or preferences. Traditional MADM techniques rely on crisp numerical values, but in many cases, precise data may not be available. Fuzzy sets allow the representation of data using linguistic terms to represent qualitative information such as high, moderate, low and membership functions that assign a degree of truth ranging between 0 and 1. This enhances the flexibility and realism of decision models. The concept of the fuzzy logic can be integrated with the MADM techniques. The alternatives and decision criteria can be evaluated using fuzzy numbers incorporating triangular or trapezoidal fuzzy function. The fuzzy operations can be applied for normalization, weighting, and ranking. In MADM, several fuzzy logic mathematical operations are employed across different phases to evaluate, compare, and rank alternatives based on multiple decision criteria. These operations help handle both quantitative and qualitative data, and some are adapted for fuzzy or probabilistic environments [4]. The following are some basic operations that is vital for applying fuzzy logic into MADM techniques.

A fuzzy set extends the concept of a classical set by allowing each element to have a degree of membership ranging between zero and one, rather than simply belonging or not belonging to the set. This flexible representation helps handle uncertainty and vagueness in various complex real-world problems. It is an extension of the classical set theory. Mathematically, a fuzzy set is a pair $(A, \mu A)$ where A is a set and μA: $A \to [0, 1]$ degree of membership. For all $x \in A$, $\mu A(x)$ is called the degree of membership of x. If degree of membership of element x, $\mu A(x) = 1$,, we say that x is fully included in (A, μA), and if $\mu A(x) = 0$,, we say that x is not included in (A, μA). If there exists some x ∈ A such that $\mu A(x) = 1$, we say that $(A, \mu A)$ is normal. Otherwise, we say that $(A, \mu A)$ is subnormal. A fuzzy set is called normal if at least one element has a membership degree of 1; otherwise, it is subnormal. To simplify representation and enhance clarity and understandability, a fuzzy set is generally denoted using a specific notation that conveniently captures its defining characteristics, such as membership values and elements [5, 6].

$$A = \frac{\mu A(x_1)}{x_1} + \ldots + \frac{\mu A(x_n)}{x_n}$$

$$A \subseteq \{x_1, x_2, ..., x_n\} = X$$

where $\frac{\mu A(x_n)}{x_n}$ is a pair degree of membership element.

1.7.1 Definitions on Fuzzy Sets

For two fuzzy sets $(A, \mu A)$ and $(B, \mu B)$, where $A, B \subseteq X$ are crisp sets, the following definitions apply to describe their relationships.

Equality: In fuzzy set theory, two fuzzy sets are said to be equal if, for every element in the universal set, their membership functions assign the same degree of membership. This ensures both sets represent identical fuzzy information. This means that if two fuzzy sets, A and B,, have the same membership value ($\mu A(x) = \mu B(x)$) for every element x in the universal set X,, then they are considered equal ($A = B$).

$A = B$ if and only if $\mu A(x) = \mu B(x)$ for $\forall x \in X$

Inclusion: A fuzzy set A is said to be included in (or a subset of) fuzzy set B, denoted as A ⊆ B, if and only if the membership value of every element in A is less than or equal to its membership value in B. Mathematically is can be defined as.

$A \subseteq B$ if and only if $\mu A(x) \leq \mu B(x)$ for $\forall x \in X$

Cardinality: In the context of fuzzy sets, cardinality is a measure of the "size" or "weight" of a fuzzy set, considering partial membership. It is not just a count of elements, but a sum of membership values, reflecting the degree to which each element belongs to the set. Mathematically is can be defined as

$$|A| = \sum_{i=1}^{n} \mu A(x_i)$$

Empty Set: In fuzzy logic, an Empty Set (also called the null fuzzy set) refers to a fuzzy set in which every element has a membership value of 0. Mathematically is can be defined as.

A is empty if and only if $\mu A(x) = 0$ for all x ∈ X.

α-Cut: In fuzzy set theory, an α-cut is a crisp, i.e., non-fuzzy subset of a fuzzy set, defined by a specific membership level (alpha). It is the set of all elements in the universal set whose membership grade in the fuzzy set is greater than or equal to the given alpha value. In simpler terms, it is like drawing a level on the fuzzy set's membership function and taking all the elements that are above that level. An alpha-cut helps to extract "crisp" versions of a fuzzy set at different levels of membership, allowing for more precise analysis and manipulation of fuzzy information. In fuzzy logic, an α-cut (alpha-cut) of a fuzzy set is a crisp (classical) set derived from a fuzzy set by extracting all elements whose membership degree is greater than or equal to a certain threshold α. Mathematically is can be defined as.

Given $\alpha \in [0, 1]$, the α-cut of A is defined by $A_\alpha = \{x \in X \,|\, \mu A(x) \geq \alpha\}$

The alpha-cut can be used in fuzzy set operations, defuzzification, and decision making which can facilitates interval-valued approximations of fuzzy sets. It may be essential for converting fuzzy sets into crisp sets for analysis or control.

1.7.2 Operations on Fuzzy Sets

In fuzzy set theory, operations on fuzzy sets extend classical set operations using membership functions. These operations handle degrees of membership between 0 and 1, making them ideal for reasoning under uncertainty or vagueness. In this section some vital operations have been discussed. Let $(A, \mu A)$ and $(B, \mu B)$ be a fuzzy set.

Complementation: The complement of a fuzzy set represents the degree to which an element does not belong to that set. It is the fuzzy logic equivalent of the classical NOT operation.

$(\neg A, \mu_{\neg \mathrm{A}})$, where $\mu_{\neg \mathrm{A}} = 1 - \mu A$

It is used in fuzzy rule negation and may be an essential operation in fuzzy set difference and logical reasoning. It helps in defining fuzzy boundaries or absence of attributes.

Height: In fuzzy set theory, the height of a fuzzy set is a measure of the maximum membership value that any element in the set has. It gives an idea of how "strongly" any element belongs to the fuzzy set.

$$h(A) = \max_{x \in X} A(x)$$

Support: In fuzzy set theory, the support of a fuzzy set refers to the set of all elements in the universe of discourse that have non-zero membership values in the fuzzy set.

$$supp(A) = \{x \in X \mid \mu A(x) \geq 0\}$$

The Support is a crisp set, not a fuzzy set. It includes all elements that have any degree of membership greater than zero. It helps identify the range or region over which the fuzzy set has influence.

Intersection: In fuzzy set theory, the intersection of two fuzzy sets represents the degree to which elements belong to both sets simultaneously. It is a generalization of the classical intersection operation.

$C = A \cap B$, where $\mu C = \min_{x \in X}\{\mu A(x), \mu B(x)\}$

Union: In fuzzy set theory, the union operation combines two fuzzy sets to determine the degree to which each element belongs to at least one of the sets. It generalizes the classical union from crisp sets.

$C = A \cap B$, where $\mu C = \max_{x \in X}\{\mu A(x), \mu B(x)\}$

Bounded Sum: In fuzzy set theory, the bounded sum operation is a method of combining the membership values of two fuzzy sets that ensures the result does not exceed a certain upper bound. It is often used to model cases where the combined membership should not exceed a maximum threshold. The bounded sum is typically defined using a specific formula that caps the value of the sum of two membership functions.

$C = A + B$, where $\mu C = \min_{x \in X}\{1, \mu A(x) + \mu B(x)\}$

Bounded Difference: The bounded difference operation in fuzzy set theory is used to compute the membership values of a fuzzy set that represent the difference between two fuzzy sets, but with an upper limit or bound. The difference operation generally describes how much an element belongs to one set but not the other. The bounded difference is particularly useful when dealing with fuzzy sets in situations where you need to subtract the membership values, but want to ensure that the result does not fall below zero or exceed a certain value.

$C = A - B$, where $\mu C = \max_{x \in X}\{0, \mu A(x) - \mu B(x)\}$

The bounded difference is useful in situations where you want to model situations where one fuzzy set "removes" or "subtracts" membership from another fuzzy set but ensure the result cannot become negative. For example, in fuzzy decision-making, it can represent the effect of subtracting one criterion from another while avoiding negative values.

Exponentiation: The exponentiation operation in fuzzy set theory is used to modify the membership values of a fuzzy set by raising them to a specified power. It helps to adjust the shape of the membership function and can be used to model different types of intensities or strengths of membership. By applying exponentiation, we can make a fuzzy set more concentrated around the highest membership value or make it more spread out. The membership values of the resulting fuzzy set have within the range [0,1].

$C = A_\alpha$ where $\mu C = (\mu A)^\alpha$ for $\alpha > 0$

By adjusting the exponentiation, control how tight or loose clusters are in a fuzzy clustering algorithm. Exponentiation can be used to tune the sensitivity of fuzzy rules and membership functions in systems like fuzzy logic controllers. The operation helps in modeling the degree of membership or strength of preference for decision-making criteria.

1.7.3 Membership Functions

A membership function defines how each element in the universe of discourse is mapped to a value between 0 and 1, representing its degree of membership in a fuzzy set. Membership functions are the backbone of fuzzy set theory and fuzzy logic-based decision-making (like fuzzy MADM), enabling the representation of vague, imprecise, or linguistic information [7, 8]. Here we will discuss some of the vital membership functions.

1.7.3.1 Triangular Membership Function

A triangular membership function is a widely used and straightforward fuzzy membership function in fuzzy logic systems. It describes how each point in the input space is assigned a membership value between 0 and 1, forming a triangular shape to represent the degree of membership. It is characterized by a triangle shape, defined by three parameters (*a*, *b*, and *c*), where '*a*' and '*c*' represent the base points of the triangle and '*b*' represents the peak. The membership value, $\mu A(x)$, ranges from 0 to 1, indicating the degree of belonging of a value '*x*' to the fuzzy set A. The triangular membership function $\mu(x : a, b, c)$ is defined as:

$$\mu(x : a, b, c) = \begin{cases} 0, & x \leq a \\ \frac{x-a}{b-a}, & a < x < b \\ \frac{c-x}{c-b}, & b < x < c \\ 0, & x > c \end{cases}$$

It is one of Simplest and intuitive shape for modeling that is often used in fuzzy inference systems like Mamdani or Sugeno. It is more suitable for real-time applications due to its low computational complexity. The area under the triangle can be interpreted as the fuzziness level or support [9].

1.7.3.2 Trapezoidal Membership Function

A trapezoidal membership function in fuzzy logic is a type of membership function that has a piecewise linear shape, resembling a trapezoid. A trapezoidal membership function is a widely used fuzzy membership function in fuzzy logic systems, especially when a flat-topped region is needed to represent full membership across a specific range of values. It is specified by four parameters: a, b, c, and d, where a and d represent the left and right endpoints, while b and c indicate the left and right support points, respectively. The function assigns input values a membership degree ranging from 0 to 1, showing how strongly an element is associated with a fuzzy set. The trapezoidal membership function $\mu(x : a, b, c, d)$ is defined as:

$$\mu(x : a, b, c, d) = \begin{cases} 0, & x \leq a \\ \frac{x-a}{b-a}, & a < x \leq b \\ 1, & b < x \leq c \\ \frac{d-x}{d-c}, & c < x \leq d \\ 0, & x > d \end{cases}$$

The shape resembles a trapezoid. The trapezoidal membership function is more flexible than the triangular function because of the flat region. It may be the Ideal when a range of values are equally representative of a fuzzy concept. It is generally used in real-world systems where precise distinctions are unnecessary over a certain interval. The trapezoidal membership function provides a flexible way to represent fuzzy sets where the boundaries are not sharply defined but have a gradual transition, often used when dealing with imprecise or vague concepts [10].

1.7.3.3 Gaussian Membership Function

A Gaussian membership function in fuzzy logic maps an input value to a membership degree between 0 and 1 using a bell-shaped curve, similar to a normal distribution. A Gaussian membership function is a smooth, bell-shaped curve used in fuzzy logic to represent uncertainty or fuzziness with continuous transitions. It is particularly useful when gradual changes are desired rather than sharp boundaries. A Gaussian membership function is defined by two parameters: namely center where the membership is maximum, i.e., 1 and the standard deviation controls the width/spread of the curve. It is defined by its center (c) and width (sigma), representing the mean and standard deviation of the Gaussian distribution, respectively. This function is commonly used in fuzzy systems for its smoothness and ease of implementation [11, 12].

$$\mu(x : c, \sigma) = \exp\left(-\frac{(x - c)^2}{2\sigma^2}\right)$$

1.8 Criteria and Attributes in Decision-Making

In decision-making, criteria and attributes serve as fundamental components that guide the evaluation, assessment, and selection of alternatives. Criteria are the standards or principles used to judge and compare each alternative. They reflect the goals, preferences, or priorities of the decision-maker and are often derived from the context of the problem. For example, in selecting the optimal and most appropriate alternative healthcare provider, criteria may include cost, quality of care, accessibility, and patient satisfaction. Attributes, on the other hand, are the measurable properties or features of each alternative corresponding to the criteria. They provide the quantitative or qualitative data needed to assess how well each alternative meets the defined criteria. For example, if cost is a criterion, the attribute could be the actual price or service fee. The effectiveness of decision-making heavily depends on clearly defining relevant criteria and identifying accurate attributes. MCDM methods use well-defined decision criteria and attributes as elements to structure complex decision problems. Proper alignment between criteria and attributes ensures objectivity, transparency, and consistency in evaluations. A well-structured framework of criteria and attributes transforms subjective preferences into rational, traceable decisions. Effectively combining criteria and attributes helps decision-makers make informed choices by systematically assessing alternatives against defined standards [13, 14].

For example, when selecting a car, criteria might include cost, fuel efficiency, safety, and performance. The corresponding attributes would be the actual price of the car, its miles per gallon rating, crash test results, and horsepower. By defining clear criteria and measuring relevant attributes, decision-makers can systematically compare alternatives. The relationship between criteria and attributes ensures structured decision-making. Criteria must be relevant, measurable, and non-redundant to avoid bias, while attributes should provide accurate and comparable data. However, poorly defined criteria or inconsistent attribute measurements can lead to unreliable rankings. Thus, a well-structured evaluation process with carefully selected criteria and objectively measured attributes is essential for making informed and rational decisions. This approach minimizes subjectivity and enhances the consistency of the final decision regarding the ranking and selection of the most optimal and appropriate alternatives [15].

1.9 Weighting and Ranking Methods

The process of solving a decision-making problem using MCDM techniques involves a structured and systematic approach composed of six distinct phases. These phases ensure that the decision-making process is both rigorous and transparent, accommodating diverse and often conflicting criteria associated with complex problems.

1.9.1 Phase 1: Problem Definition and Construction of the Decision Matrix

The initial step in any MCDM methodology is to clearly define the decision problem. This includes systematically identifying the objectives of the decision, the relevant criteria or attributes that will influence the decision, and the possible alternatives or options available. A decision matrix is then constructed to represent this information. The matrix displays alternatives as rows and criteria as columns, with the elements of the matrix representing the performance or score of each alternative under each criterion. This phase is critical, as the quality of the final decision is directly influenced by the clarity and completeness of the initial problem formulation.

1.9.2 Phase 2: Normalization of Decision Criteria

In most real-world scenarios, the criteria involved are expressed in different units or scales, such as cost in dollars, time in hours, or efficiency in percentages. To ensure comparability, it is necessary to normalize the decision matrix. Normalization transforms different scales into a dimensionless and comparable format without altering the fundamental meaning of the data. This step ensures that no single criterion dominates the decision due to its scale alone. Techniques such as min–max normalization, vector normalization, or logarithmic scaling transform criteria values into a uniform range (generally 0–1). This step ensures compatibility among diverse criteria, preventing bias due to varying measurement scales. The detailed discussion of various normalization methods is provided in Chapters 2.

1.9.3 Phase 3: Weighting of Criteria

Once the data is normalized, the relative importance of each criterion is assessed. Weights are assigned to each criterion based on expert judgment, stakeholder input, or mathematical techniques such as entropy methods or pairwise comparisons. These weights reflect the significance of each criterion in the context of the decision problem. Accurate and meaningful weighting is essential, as it directly influences the ranking of the alternatives. Researchers have proposed several methods to assign weight to each decision criterion. Researchers have grouped these weighting approaches to assigning weight to the criteria to each decision criteria into three categories: subjective, objective, and hybrid methods. The weight of the criteria for each decision criterion reflects the amount of information to assess the performance score of each alternative. The subjective approach of assigning significance to each decision criterion is either difficult or unreliable due to the decision-maker's reliance on their knowledge, experience, and expertise. The objective weight method is a more straightforward method without the intervention of any decision-maker. In the objective weight method, each attribute is taken as an information source, and the weight of significance of each criterion reflects the amount of information contained in each of them. The hybrid weighting methods have been proposed. The detailed discussion of weighting methods is provided in Chapters 3, 4, and 5.

1.9.4 Phase 4: Scoring and Ranking of Alternatives

In this phase, the weighted normalized values are aggregated to calculate a performance score for each alternative. Different MCDM techniques, such as the Analytic Hierarchy Process (AHP), Technique for Order Preference by Similarity to Ideal Solution (TOPSIS), or Weighted Sum Model (WSM), may be employed to compute these scores. Alternatives are then ranked based on their performance scores, with the highest-scoring option considered the most preferred solution. The detailed discussion of weighting methods is provided in Chapters 6.

1.9.5 Phase 5: Validation, Sensitivity Analysis, and Applicability in Different Domain of Application

After obtaining the rankings, it is vital to validate the robustness and consistency of the results. Sensitivity analysis is commonly used to determine how changes in weights or criteria affect the outcome. Additionally, applying multiple MCDM techniques can help verify whether the rankings are consistent across methods. This step enhances the reliability and defensibility of the decision-making process. The MCDM techniques are an optimization technique that has the potential to address complex decision problems in science, engineering, and economics where diverse, complex, and conflicting decision criteria can be involved. Recently MCDM methods are increasingly prevalent and effective for addressing complex decision problems, including economics, engineering, science, medicine, education, social science, and many other emerging domains. The detailed discussion on the applicability of MCDM techniques across various application domains is provided in Chap. 7.

1.9.6 Phase 6: Limitations and Future Research Directions

The final phase involves a reflective assessment of the MCDM process, identifying limitations such as uncertainty in data, subjectivity in weighting, and the scalability of methods. Furthermore, this phase looks toward the future, proposing areas for further research, such as integrating artificial intelligence, handling dynamic criteria, or improving stakeholder engagement. Advancing MCDM methodologies will continue to play a crucial role in solving increasingly complex decision problems in various domains. Chapter 8 provides an in-depth discussion on the challenges faced in MCDM techniques and explores emerging research trends, highlighting potential directions for future advancements and methodological improvements.

Weighting and ranking methods are core components of MCDM techniques. Weighting involves assigning relative importance to each criterion, while ranking methods evaluate and order the alternatives based on performance scores. Together, they guide decision-makers in identifying the most suitable option among multiple conflicting criteria and alternatives. The subsequent chapters of the book have deep discussion on each phase of the MCDM techniques mentioned above.

1.10 Summary

Decision theory involves the use of mathematical models, principles, and methods to support individuals, organizations, and policymakers in analyzing decisions, assessing potential outcomes, and selecting the most appropriate course of action based on preferences, probabilities, and consequences. This chapter begins with a general overview aligned with key decision-making concepts and then delves into specific components essential to understanding decision theory. The sections include the basics of decision theory, single-attribute decision models, and multi-attribute decision-making. It also introduces the fundamentals of fuzzy theory, examines the significance of criteria and attributes in decision-making, and concludes with a brief discussion on MCDM techniques, highlighting two core components: weighting and ranking methods.

References

1. D.W. North, A tutorial introduction to decision theory. IEEE Trans. Syst. Sci. Cybern. **4**(3), 200–210 (1968)
2. G. Parmigiani, L. Inoue, *Decision theory: principles and approaches.* (Wiley, 2009).
3. E.K. Zavadskas, Z. Turskis, S. Kildienė, State of art surveys of overviews on MCDM/MADM methods. Technol. Econ. Dev. Econ. **20**(1), 165–179 (2014). https://doi.org/10.3846/20294913.2014.892037
4. C. Kahraman, Multi-criteria decision making methods and fuzzy sets, in *Fuzzy Multi-Criteria Decision Making*, vol. 16, ed. by C. Kahraman (Springer Optimization and Its Applications, Boston, MA; Springer US, 2008), pp. 1–18. https://doi.org/10.1007/978-0-387-76813-7_1
5. J. J. Buckley, E. Eslami, *An introduction to fuzzy logic and fuzzy sets*, vol. 13. (Springer Science & Business Media, 2002), https://books.google.co.in/books?hl=en&lr=&id=IzN93YlrKlwC&oi=fnd&pg=PA1&dq=Fuzzy+logic+basic+theory+and+operation&ots=YzYbSNkGE7&sig=z31FaOO_Z2iKl87tlNUHyjbNcxE. Accessed: 09 May 2025
6. R. A. Aliev, *Fundamentals of the fuzzy logic-based generalized theory of decisions.* (Springer, 2013), https://books.google.co.in/books?hl=en&lr=&id=uo-6BQAAQBAJ&oi=fnd&pg=PP2&dq=Fuzzy+logic+basic+theory+and+operation&ots=KgiLPqHXK_&sig=obP_hxUoVkdbiNLnLsEz7ORhggA. Accessed 09 May 2025
7. N.D. Singpurwalla, J.M. Booker, Membership functions and probability measures of fuzzy sets. J. Am. Stat. Assoc. **99**, 467. Accessed 20 May 2025. https://doi.org/10.1198/016214504000001196
8. J. Dombi, Membership function as an evaluation. Fuzzy Sets Syst. **35**(1), 1–21 (1990). https://doi.org/10.1016/0165-0114(90)90014-W
9. W. Pedrycz, Why triangular membership functions? Fuzzy Sets Syst. **64**(1), 21–30 (1994). https://doi.org/10.1016/0165-0114(94)90003-5
10. A. Barua, L. Mudunuri, O. Kosheleva, Why trapezoidal and triangular membership functions work so well: towards a theoretical explanation, *Departmental Technical Reports (CS)*, (2013). https://scholarworks.utep.edu/cs_techrep/783
11. V. Kreinovich, C. Quintana, L. Reznik, Gaussian membership functions are most adequate in representing uncertainty in measurements, in *Proceedings of NAFIPS* (1992), pp. 15–17, https://www.researchgate.net/profile/Vladik-Kreinovich/publication/23894056_Gaussian_Membership_Functions_Are_Most_Adequate_in_Representing_Uncertainty_in/links/0deec5231f36f43731000000/Gaussian-Membership-Functions-Are-Most-Adequate-in-Representing-Uncertainty-in.pdf. Accessed 20 May 2025
12. Y. Li, J.-M. Deng, M.-Y. Wei, Meaning and precision of adaptive fuzzy systems with Gaussian-type membership functions. Fuzzy Sets Syst. **127**(1), 85–97 (2002). https://doi.org/10.1016/S0165-0114(01)00154-3
13. G.-H. Tzeng, J.-J. Huang, *Multiple attribute decision making: methods and applications.* (CRC press, 2011)
14. R. V. Rao, *Introduction to Multiple attribute decision-making (MADM) methods.* SpringerLink.https://doi.org/10.1007/978-1-84628-819-7_3. Accessed 20 May 2025

15. K.P. Yoon, C.-L. Hwang, *Multiple attribute decision making: an introduction.* (SAGE Publications, 1995)

Chapter 2
Normalization Strategies

Abstract Normalization is the first step of any MADM approach. It is used to standardize data and eliminate redundancies, inconsistencies, and anomalies by adjusting values to fall within a predetermined range or scale. In this chapter, we will explore importance, purpose, step-by-step computation, types, advantages, disadvantages, and applications in different areas of the domain of different approaches to normalization. This chapter also has a discussion of utility theory and its different types that is geared toward facilitating effective decision-making processes.

Keywords Normalization · Benefit criteria · Cost criteria · z-score · Logarithmic normalization · Linear normalization · Vector normalization · Max–min normalization. Sum normalization · MADM

2.1 Importance of Normalization

The aim of the MADM approach is to rate and prioritize a set of alternatives that best satisfy a given set of decision criteria, which may be complex and conflicting. Criteria are a set of requirements or independent attributes that have to be satisfied by several alternatives. Each decision criterion of an alternative may be measured in different units such as degrees, kilograms, and meters. Each decision criterion may have a distinct range of values, varying across different alternatives, which can influence the evaluation process and the final selection in multi-criteria decision-making. To ensure all decision criteria for the alternatives are comparable and compatible, their values must be normalized to a common scale. This common and dimensionless value of each decision criterion of all alternatives enables meaningful aggregation into a unified evaluation score. Hence, normalization is an essential part of any decision-making process, as it converts input data into a standardized format, enabling the application of MADM methods for ranking and selecting the best alternatives. There are several definitions for data normalization, depending on the study domain. In context of databases management system, normalization is the process of reducing redundancy. It refers to organizing data attributes within a relational data

A. K. Yadav et al., *Multi-attribute Decision-Making: Bridging Theory and Practice*,
Transactions on Computer Systems and Networks,
https://doi.org/10.1007/978-981-95-4782-1_2

model into structured tables to enhance cohesion and improve data management efficiency. In statistics, normalization typically means adjusting values measured on different scales and in different units to a common scale, handling missing or default value through appropriate statistical values often before performing aggregation or averaging. There are many other definitions of normalization that exist, depending on the context or study domain. This chapter concentrates on normalization techniques within the framework of MADM. In general, normalization in MADM is a transformation process that converts different criteria into a common scale to make them comparable. Once the decision matrix is defined, it undergoes pre-processing before directly computing the performance scores of alternatives. This step ensures the comparability of all decision criteria, making the data suitable for effective decision modeling. In MADM, normalization techniques are commonly used to convert criteria with different measurement units and a diverse range into a predefined scale, generally within the [0–1] range. Various research studies have indicated that the selection of a normalization technique can significantly impact the ranking of alternatives. Researchers have shown that some MADM methods may be more appropriate for specific decision-making problems and approaches than others [1]. The proper selection of attributes and alternatives are key points in the MADM model. At first look, the problem does not seem to be mathematically constrained. But, in fact, the selection of criteria and alternatives are the features constraining the model implicitly. If the selection of attributes and alternatives is not carefully made, the solving algorithm can yield fatal errors. During a decision path, engineers must select criteria with superior performance while maintaining minimal interactions between them [2]. MADM methods use normalization techniques to aggregate criteria with numerical and comparable data. With the rise of cyber-physical systems that gather huge amounts of data from diverse sensors and sources, choosing the right normalization technique to normalize this data in the context of applying machine learning and deep learning has become a challenge. Therefore, data integration and criteria aggregation are similar processes, combining values from either criteria or sensors to obtain a common score. This chapter has a comprehensive discussion on metrics for assessing the most appropriate normalization techniques for decision problems.

2.2 Purpose of Normalization

Normalization is one of the crucial techniques in the MADM approach for decision-making and data analysis to eliminate scale disparities among criteria measured in different measurable units, enabling fair comparisons. By converting data to a common scale, it facilitates meaningful trade-offs and precise weighting in multi-criteria decision-making. It helps maintain value ranges for interpretability and reduces the risk of rank reversal by preserving proportional relationships. Normalization ensures that no single criterion dominates due to its scale, supporting stable and rational outcomes. It also enhances MADM algorithm performance by promoting a

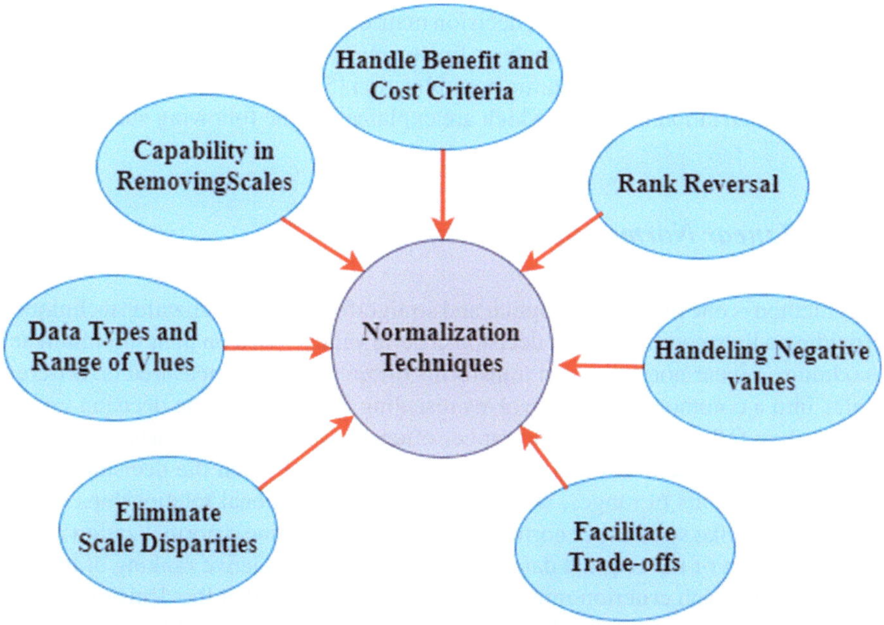

Fig. 2.1 Purpose of normalization techniques

reduction in rank abnormality, faster convergence, and numerical stability. Normalization is one of the vital aspects of the MADM for consistency, fairness, and clarity in analytical processes. Following are the core aspects of normalizing decision-making. Figure 2.1 demonstrates the different purposes fulfilled by the normalization techniques for obtaining stable and consistent rankings of alternatives to the given decision problem.

2.3 Normalization Methods

Normalization is a fundamental first step in any decision-making process, used to convert data with different units into a common scale, making all values comparable. It is a crucial step in any decision-making process to evaluate alternatives and rank them. Normalization changes different types of data, such as opinions, numbers, and various units, into a standard numerical format that can be compared, allowing for the combination of criteria to assess the options for decision-making [3]. There are some critical issues in identifying the appropriate number and effective criteria of alternatives from available alternatives. It may be due to different characteristics and a variety of measurable dimensions. This may lead to comparison incompatibility. This is the initial stage in almost every MADM technique and is also known as standardization. It is useful to convert various measurable unit values in the range of zero and one [4]. Normalization of decision matrices can be calculated using different approaches of normalization, such as linear, max–min linear, sum linear, vector-enhanced accuracy, and logarithmic normalization [5]. We have covered some of the crucial normalization methods, which are explained in the following subsections.

2.3.1 Linear Normalization

This method is one of the most popular and straightforward normalization techniques, providing a linear adjustment of decision criteria values based on their minimum and maximum. Linear normalization transforms different criteria measured on different scales into a common scale. It involves rescaling the values of the decision criteria so that they fall within a specific range, often between 0 and 1, or sometimes − 1 to 1. Linear normalization is particularly more useful when the decision criteria have different units or ranges, as it preserves the proportional relationships among the values. Unlike some other normalization techniques, linear normalization retains the distribution of the original data. This means that the relative ranking of alternatives based on each criterion remains consistent, ensuring that the decision-making process is fair and meaningful. It is versatile and can be applied to various types of data, including both quantitative and qualitative decision criteria. This capability of this normalization technique making it more appropriate for a wide range of decision-making scenarios in MADM. This normalization technique is very simple and suitable for aggregation methods such as SAW.

$$r_{i,j} = \begin{cases} \frac{x_{i,j}}{x_j^{\max}} \ \forall c_j \in AT_1 \\ \frac{x_j^{\min}}{x_{i,j}} \ \forall c_j \in AT_2 \end{cases}$$

The characteristics such as simplicity and efficiency, preservation of data distribution, flexibility for different applications, and better comparability make linear normalization a widely used technique in MADM. It is a more applicable technique, especially when a decision-maker needs a straightforward approach that does not overcomplicate the decision-making process. But on other hand linear normalization relies on the minimum and maximum values of the data, it can be sensitive to extreme values, which may distort the normalization process. This sensitivity to the extreme values may lead to outliers that can distort the scale, leading to misrepresentation of the data.

2.3.2 Max–Min Linear Normalization

It is a linear transformation method to standardize diverse decision criteria measured on different scales into a common range. Which ensures the minimum value of the criterion is transformed to 0, the maximum value to 1, and the other values proportionally scaled in between. This normalization is generally used in SAW, VIKOR, and PROMETHEE for scaling to [0, 1].

$$r_{i,j} = \begin{cases} \frac{x_{i,j} - x_j^{\max}}{x_j^{\max} - x_j^{\min}} & \forall\, c_j \in AT_1 \\ \frac{x_j^{\max} - x_{i,j}}{x_j^{\max} - x_j^{\min}} & \forall\, c_j \in AT_2 \end{cases}$$

Max–min normalization method is beneficial because it does not alter the relative ordering of the alternatives but simply scales them in a manner that removes the effect of varying units or scales across different attributes. The max–min normalization is less sensitive to the extreme values of the criteria. In some cases, when the range of values for a decision criterion is not linear, max–min Liner Normalization may still help in making the data comparable by scaling it within the same interval, although the relationship between attributes remains non-linear. This capability makes it useful for comparing attributes that do not follow a straight-line relationship. Min–max normalization, despite its ability to preserve relative comparison, adapt to different scales, offer simplicity and ease of use, remain robust to outliers, and be effective for nonlinear relationships, faces the problem of sensitivity to range. This sensitivity to range is possible in min–max normalization due to the dependence on the minimum and maximum values in the data. If the range of the data is very small, it might lead to values being clustered around a narrow range, potentially making it harder to distinguish between the alternatives.

2.3.3 Sum Linear Normalization

It involves scaling the values of each decision criterion so that they collectively sum up to a predetermined value, often 1 or 100, ensuring that their combined weights align with the total weight distributed to all criteria. It also ensures that the relative importance of each criterion to the overall decision is reflected accurately.

$$r_{i,j} = \begin{cases} \frac{x_{i,j}}{\sum_{i=1}^{i=n} x_{i,j}} & \forall\, c_j \in AT_1 \\ \frac{\frac{1}{x_{i,j}}}{\sum_{i=1}^{i=n} \frac{1}{x_{i,j}}} & \forall\, c_j \in AT_2 \end{cases}$$

The sum linear normalization scales all decision criteria linearly to a common range [0, 1] by dividing each value by the sum of all values for that decision criterion. Unlike min–max normalization, which uses extreme values, it considers the entire dataset. The ratios between normalized values remain the same as in the original data. It is based on the principle of the sum, unlike min–max, which is bounded by min/max or mean-centered or Z-score normalization techniques. Sum linear normalization is a more suitable alternative technique when relative proportions matter more than absolute bounds. The sum linear normalization is simple and intuitive, preserves proportionality, and is suitable for MCDM, economics, and data science. But similar to some other normalization techniques, it is also sensitive to extreme values. A very large value in a criterion can shrink other normalized values close to zero. If a decision matrix or data set has fewer observations, the sum can be highly influenced by a single entry, leading to skewed normalization. The process of this normalization assumes all values are positive. The negative numbers can lead to meaningless results, such as negative sums. Therefore, this method is only suitable for positive values. In methods like Euclidean distance, such as clustering, it is not ideal because it distorts the data geometry. So, it is also not a better alternative when preserving distance or variance is important. Linear sum-based normalization acquired less time and space than other normalization techniques. Because linear sum-based normalization requires less time and space, it is more suitable for TOPSIS decision-making to achieve the optimal solution [5].

2.3.4 *Vector Normalization*

It is also known as Euclidean normalization. The core principle of vector normalization is based on the computation of distances from the ideal and anti-ideal solutions, which ensures that all decision criteria contribute equally, regardless of original units. It transforms the original vector values so that they maintain their relative proportions but are adjusted to adhere to a certain norm, usually a magnitude of 1 for ease of comparison. Its principle is based on the Euclidean norm, which ensures that the vectors maintain their relative direction and proportionality but are adjusted to have a standardized magnitude, enabling fair comparisons and collective analysis. This normalization technique is more appropriate for MADM methods like TOPSIS, where Euclidean distances to ideal solutions are calculated.

$$r_{i,j} = \begin{cases} \frac{x_{i,j}}{\sqrt{\sum_{i=1}^{m} x_{i,j}^2}} & \forall c_j \in AT_1 \\ 1 - \frac{x_{i,j}}{\sqrt{\sum_{i=1}^{m} x_{i,j}^2}} & \forall c_j \in AT_2 \end{cases}$$

After normalization, each decision criteria contribution is viewed as a proportion of total magnitude, which makes it more useful in multi-criteria ranking. It is one of the most widely used normalization techniques in TOPSIS. Unlike sum liner normalization, it has the capability to handle both benefit and cost criteria with proper post-normalization adjustments. If criteria relationships follow an exponential or logarithmic scale, this technique may be the better choice of normalization. If one value in the vector is much larger than others, it dominates the normalization. It requires adjustments if data contains negative values in the decision matrix or data set. It can have division instability due to zero- or near-zero values in the data set.

2.3.5 Enhanced Accuracy Normalization

Enhanced accuracy normalization is one of the recent and refined normalization techniques used in MADM with the objective of addressing limitations in traditional normalization such as sum linear normalization, min–max, and vector normalization.

$$r_{i,j} = \begin{cases} 1 - \frac{x_j^{\max} - x_{i,j}}{\sum_{i=1}^{i=n} x_j^{\max} - x_{i,j}} & \forall c_j \in AT_1 \\ 1 - \frac{x_{i,j} - x_j^{\min}}{\sum_{i=1}^{i=n} x_{i,j} - x_j^{\min}} & \forall c_j \in AT_2 \end{cases}$$

It reduces the distortion effect that can occur with extreme values or irregular data distributions.

2.3.6 Logarithmic Normalization

The principle of this method is to apply a logarithmic function to the original values to rescale them into a comparable format. This method is more useful when the data spans a wide range and contains outliers or extremely large values, helping to manage and normalize such variations. This method is more effective in compressing a wide range of values, making extreme values more manageable while retaining the relative proportions and differences between data points.

$$r_{i,j} = \begin{cases} 1 - \frac{\ln(x_{i,j})}{\ln\left(\prod_{i=1}^{i=m} x_{i,j}\right)} & \forall c_j \in AT_1 \\ 1 - \frac{\frac{\ln(x_{i,j})}{\ln\left(\prod_{i=1}^{i=m} x_{i,j}\right)}}{m-1} & \forall c_j \in AT_2 \end{cases}$$

where $1 \leq i \leq m$ and $1 \leq j \leq n$. $x_j^{\max}$ and $x_j^{\min}$ is the maximum and minimum value of jth attribute, respectively. Logarithmic normalization is a relatively newer approach in the context of MADM. Unlike traditional linear approaches such as min–max scaling or vector normalization, it is based on a logarithmic function. This transformation helps to compress the scale of the data, particularly useful when dealing with criteria that have a wide range of values or are heavily skewed. The use of the logarithmic function compresses the larger values more significantly than the smaller ones, reducing the dominance of extreme values. This normalization technique reduces the skewness in the data distribution of the criteria, making it more symmetrical. Logarithmic normalization offers an interesting alternative to traditional normalization techniques in MADM, particularly when dealing with data that has a wide range or is significantly skewed. Logarithmic normalization is particularly beneficial in MADM when dealing with criteria that exhibit large extremes.

The standard logarithmic function is undefined for zero and negative numbers. Therefore, if the original data contains such values, adjustments such as adding a small positive constant are necessary before applying the logarithmic transformation. The choice of this constant can potentially influence the normalized results. The logarithmic transformation is non-linear, which can alter the linear relationships present in the original data. This might be a concern depending on the specific MADM method being used and the underlying assumptions. Therefore, a careful consideration of its applicability, especially concerning zero and negative values and the interpretability of the results, is essential. This normalization technique works well with entropy-based weighting, where normalized values are used to calculate diversity or information gain across criteria. It is a perfect choice for data that follows a power law or exponential scale. Modified TOPSIS, fuzzy MCDM, and hybrid MADM methods, which critically depend on data preprocessing, have increasingly utilized It [6–8].

2.3.7 Z-Score Normalization

This normalization technique is not very common in MADM but useful for handling outliers in machine learning during preprocessing of data.

$$r_{ij} = \frac{x_{ij} - \mu_j}{\sigma_j}$$

All the MADM methods work in two phases. In the first phase, MADM uses various methods for alternative criteria priority weighting. After computing and allocating appropriate weight to alternative criteria, MADM goes into the second phage. In the second phase, it uses various alternative ranking methods to rank the alternatives and select the best of them. In the next section, we will talk about more Network Criteria Priority Weight Allocation Methods and then discuss different ways to rank them [9, 10]. A comprehensive, precise comparison summary of all the above-discussed normalization techniques has been given in Table 2.1.

Table 2.1 Summary of comparison of normalization techniques

Method/ feature	Linear normalization	Max–min linear normalization	Sum linear normalization	Vector normalization	Enhanced accuracy normalization	Logarithmic normalization	Z-Score normalization
Formula	$r_{i,j} = \begin{cases} \frac{x_{i,j}}{x_j^{\max}} & \forall c_j \in AT_1 \\ \frac{x_j^{\min}}{x_{i,j}} & \forall c_j \in AT_2 \end{cases}$	$r_{i,j} = \begin{cases} \frac{x_{i,j}-x_j^{\max}}{x_j^{\max}-x_j^{\min}} & \forall c_j \in AT_1 \\ \frac{x_j^{\max}-x_{i,j}}{x_j^{\max}-x_j^{\min}} & \forall c_j \in AT_2 \end{cases}$	$r_{i,j} = \begin{cases} \frac{x_{i,j}}{\sum_{i=1}^{i=n} x_{i,j}} & \forall c_j \in AT_1 \\ \frac{\frac{1}{x_{i,j}}}{\sum_{i=1}^{i=n} \frac{1}{x_{i,j}}} & \forall c_j \in AT_2 \end{cases}$	$r_{i,j} = \begin{cases} \frac{x_{i,j}}{\sqrt{\sum_{i=1}^{m} x_{i,j}^2}} & \forall c_j \in AT_1 \\ 1 - \frac{x_{i,j}}{\sqrt{\sum_{i=1}^{m} x_{i,j}^2}} & \forall c_j \in AT_2 \end{cases}$	$r_{i,j} = \begin{cases} 1 - \frac{x_j^{\max}-x_{i,j}}{\sum_{i=1}^{i=n} x_j^{\max}-x_{i,j}} & \forall c_j \in AT_1 \\ 1 - \frac{x_{i,j}-x_j^{\min}}{\sum_{i=1}^{i=n} x_{i,j}-x_j^{\min}} & \forall c_j \in AT_2 \end{cases}$	$r_{i,j} = \begin{cases} 1 - \frac{\ln(x_{i,j})}{\ln\left(\prod_{i=1}^{i=m} x_{i,j}\right)} & \forall c_j \in AT_1 \\ 1 - \frac{\frac{\ln(x_{i,j})}{\ln(\prod_{i=1}^{i=m} x_{i,j})}}{m-1} & \forall c_j \in AT_2 \end{cases}$	$r_{ij} = \frac{x_{ij}-\mu_j}{\sigma_j}$
Data distribution	Uniform or linearly distributed	Linear, decreasing values	Any	Scaled, geometric or angle-based data	Non-linear, real-world skewed data	Exponential, skewed, multiplicative data	Normal/ Gaussian distribution
Sensitivity to Extream	High	High	Moderate	High	Low	Low	Moderate
Sum of normalized values	Not fixed	Not fixed	1 (column-wise)	√(sum of squares) → normalizes magnitude	Not fixed	1 (column-wise)	Not fixed
Range of normalized values	[0, 1]	[0, 1]	(0, 1)	(0, 1)	[0, 1] or customized	(0, 1)	Typically (−3, +3)
Advantages	Simple, intuitive and preserves ratios	Handles cost criteria directly, Bounded output	Simple implementation and Preserves proportionality	Preserves direction; used in TOPSIS	Reduces ranking errors; robust to noise, reduces bias, optimizes weights	Compresses wide ranges; mitigates outliers, Handles exponential-scale data	Captures variability; Robust to extremes, preserves distribution
Disadvantages	May produce negative values, Skewed by large extremes	Undefined for 0 or negative values, Distorted by extreme values	May need fine-tuning due to Highly sensitive to extremes	Sensitive to extremes, fails if a vector has zero magnitude	Does not preserve relative differences well and computationally intensive	Sensitive to extremes, fails for zero/negative values without ε	Affected by extreme values, no fixed bounds, loses original scale
Best use case	When data ranges vary significantly	MADM with mixed (cost/benefit) attributes	Weighted decision models (e.g., SAW)	Cosine similarity, TOPSIS	Healthcare, IoT, dynamic MADM	When values span several orders of magnitude	Statistical or probabilistic analysis

Studies comparing MADM results obtained using traditional normalization methods versus Logarithmic Normalization have shown that the final rankings of alternatives can differ. This highlights the sensitivity of MADM outcomes to the choice of normalization technique and the importance of selecting a method appropriate for the data and the decision context [11]. Which normalization technique is the most effective and it is difficult to deduce directly? It depends on the nature of the problem and the algorithm used to select the best alternative.

2.4 Application of Normalization

Normalization plays a pivotal role in MADM techniques by converting decision criteria measured on different units, ranges, and scales into a common dimensionless format. This standardization is essential for accurately comparing and aggregating diverse criteria, ensuring that each contributes appropriately to the decision-making process. Different MADM methods may respond variably to the chosen normalization technique. For instance, methods like TOPSIS and VIKOR are sensitive to the normalization approach, which can influence the final ranking of alternatives. Therefore, selecting an appropriate normalization method aligned with the specific MADM technique and the nature of the data is crucial for reliable decision-making. Normalization is crucial in MADM as it standardizes diverse criteria units, enabling fair comparisons and aggregation. By converting varied measurements into a common scale, it ensures that no single criterion disproportionately influences the decision outcome. Moreover, the choice of normalization technique can significantly impact the ranking of alternatives; inappropriate methods may lead to rank reversal, altering preferences without actual performance differences.

2.5 Utility Function

The decision-making process seeks to make a purposeful selection from among a set of alternatives in light of a given objective and preferences. Utility theory is based on preference relationships since each preference is represented numerically through the use of a utility function that associates a value for each alternative. The process then consists of a comparison among different alternatives. The utility theory assumes that any decision is made on the basis of the utility maximization principle, according to which the best choice is the one that provides the highest utility to the decision-maker. When utility is evaluated based on a single criterion, it is referred to as a single-criterion utility function. In contrast, when utility is assessed across multiple criteria, it is known as a multi-criteria utility function, typically derived by aggregating several single-criterion utility functions. A single-criterion utility function

$U(x)$ assigns a value to each criterion x based on user preferences through a mathematical expression. As this function is typically used for maximization, it is important to distinguish between two types of criteria: upward criteria (e.g., bandwidth or signal strength), where higher values are preferred, and downward criteria (e.g., energy consumption or cost), where lower values are more desirable. For downward criteria, the utility is inversely proportional to the criterion value.

The basic utility theory was developed by Von Neumann and Morgenstern [12]. Since its inception, the theory has evolved through various advancements and interpretations. The ability of a good or service to satisfy a human need is referred to as utility in microeconomics. A related concept is the utility function, which represents the satisfaction a consumer derives from a particular good or service. Because individual preferences vary, different consumers may assign different utility values to the same product. Therefore, personal preferences must be considered when evaluating utility. This concept of utility is applicable to both single-criterion (attribute or characteristic) and multi-criteria outcomes. A utility function is defined mathematically as a function $U(w, x)$ from a set of observed product criteria x (by the user) and the user preferences w into a real number. Since the user preferences for a set of criteria stay the same when looking at different options, we will just call $U(x)$ the utility function related to the criteria vector x of the product in question. We further describe the mathematical properties of such a function. Utility is an ordinal concept. It is a means to quantify the preferences among alternatives when making decisions. Such a preference relation can be represented by a continuous utility function. The fundamental assumption in the utility theory is that the decision-maker always chooses the alternative with the highest utility value; that is, the decider is assumed to be rational. Note also that just knowing that a user prefers good p to good q gives us no indication of how much more he prefers p to q. If $U\left(x_p\right) = 3U\left(x_q\right)$, the user prefers p over q, but p is not necessarily three times better than q. In the alternative selection, when evaluating the utility of an alternative, we distinguish between the upward and downward criteria associated with an alternative. The users prefer the higher values of upward criteria and lower values of downward criteria.

In the following, we investigate the utility form for the upward criteria and then extend the results to the downward criteria. Consider a utility function $U(x_i)$ of an upward quality related parameter (x_i), $0 \leq x_i < \infty$. Without loss of generality, we can consider x_i as a number of resources that an access network can allocate to the user. As an upper and/or lower bound will always exist for every parameter due to technological constraints and/or user's requirements that is $\left(x_\alpha \leq x_i \leq x_\beta\right)$, the utility has also an upper bound value. Consequently, we can normalize the utility to scale the interval [0,1], that is, $U(x_i) \in [0, 1]$. Firstly, the utility function should be twice differentiable on interval $[x_\alpha, x_\beta]$. This reflects the fact that the utility level should not change drastically given a very small change to the criterion value, and the marginal utility should be regular. Secondly, the utility function is a non-decreasing function of x_i; that is the more resources are allocated to the user, the higher the user's utility. However, when the allocated resources reach a certain threshold and user satisfaction reaches a high level, the utility improvement vanishes. This phenomenon effectively illustrates the concept of diminishing marginal utility that is $\lim_{x_\alpha \longrightarrow x_\beta} U(x_i) = 0$. The effect of diminishing marginal utility implies the concavity of $U(x_i)$ for x_i greater than a given value. The above requirements can be summarized in the following conditions statements:

$$U(x_i) \geq 0 \tag{2.1}$$

$$\exists x_c : U(x_i) < 0, \ \forall x_i \geq x_c \tag{2.2}$$

Similarly, when a quality-related parameter x_i falls below a certain threshold and the utility approaches zero, users become indifferent to further decreases in x_i. In other words, if the allocated resource remains below the minimum required level, increasing it yields negligible improvement in utility. This behavior indicates that the utility function $U(x_i)$ is convex for values of x_i below a certain threshold.

$$\exists x_v : U(x_i) > 0, \forall x_i \leq x_v \tag{2.3}$$

For an upward criterion, the utility function is defined by its twice differentiability along with conditions (2.1), (2.2) and (2.3). In contrast, for a downward criterion x_i the corresponding utility function $U(x_i)$ should satisfy the following properties:

(i) It must be twice differentiable,
(ii) It should decrease with increasing x_i,
(iii) It should be concave when x_i is below a certain threshold, and
(iv) It convex when x_j exceeds another threshold. Based on these properties, the form of $V\left(x_j\right)$ can be conveniently expressed as $V\left(x_j\right) = 1 - U\left(x_j\right)$

In addition to its functional characteristics, it is important to note that a utility function is unique up to a monotonic transformation. This means that any monotonic transformation of a utility function yields another utility function that represents the same underlying preference relation. More generally, if $g(.)$ is an increasing function and $u(.)$ is a utility function, $g\,o\,u = g(u(.))$ is a utility function having the same reference relation as $u(.)$. Numerous studies in the literature have explored various forms of utility functions, including step, piecewise, logarithmic, exponential, and sigmoidal types. Researchers have also designed utility functions to capture the elasticity of running applications or the user's frustration in response to service degradation. However, despite the wide range of proposed utility forms, there remains no clear guidance on which type is most appropriate for a given context. The extensive discourse on utility functions falls outside the purview of this book. Table 2.2 has mathematical formulas for some vital utility functions.

Table 2.2 Utility function with required conditions

Utility function	Generalized mathematical formula	Increasing and differentiability	Concavity	Convexity
Linear piecewise	$U(x) = \begin{cases} 0, & x < x_{\min} \\ \frac{x - x_{\min}}{x_{\max} - x_{\min}}, & x_{\min} \le x \le x_{\max} \\ 1, & \text{Otherwise} \end{cases}$	Yes	No	No
Logarithm	$U(x) = \ln(x)$ or $U(x) = \ln(1 = ax),(a > 0)$	Yes	Yes	No
Exponential	$U_1(x) = e^{(x-M)}$, $(0 \le x \le M)$, $U_2(x) = 1 - e^{ax},(a > 0)$	Yes	Yes	Yes
Sigmoid	$U_1(x) = \frac{1}{1+e^{C(x_m - x)}}$, $(C, x_m > 0)$ $U_2(x) = \frac{\left(\frac{x}{x_m}\right)^C}{1+\left(\frac{x}{x_m}\right)^C}$, $(x_m > 0,\ C \ge 2)$	Yes	Yes	Yes

The table also highlights essential properties that a valid utility function must satisfy: twice differentiability, monotonicity, and either concavity or convexity. A utility function is deemed appropriate only when these criteria are met, enabling the identification of an optimal point based on the defined utility model.

2.6 Challenges in Normalization

A significant current focus in data analysis is identifying the most appropriate normalization techniques in MCDM methods used in complex decision problems. This process is especially crucial in complex collaborative decision-making systems that handle a wide range of heterogeneous data sources. Using different normalization techniques may result in different rankings of alternatives. Therefore, choosing the most appropriate normalization technique for each MADM problem enhances the accuracy of alternative rankings. Despite numerous efforts in this direction, challenges persist due to a lack of coherence and the absence of a robust evaluation framework. This problem has led researchers to suggest a new way to assess different normalization techniques in MADM problems, especially when choosing partners or suppliers in collaborative networks, using various metrics. The literature proposes several normalization techniques, and specific MADM methods utilize some of these.

For instance, the TOPSIS method implements the vector normalization technique, while AHP utilizes the sum linear normalization for obtaining consistent ranking. But there is no consensus among researchers about the conditions for utilizing each specific normalization technique. Several papers have paid attention to the role and importance of normalization techniques in MADM methods. They pointed out that using different normalization techniques may lead to different rankings of alternatives and cause a degradation in the accuracy of the final solution. This situation prompted us to investigate the compatibility of normalization techniques with MADM methods [13, 14].

There are several challenges in selecting and applying normalization techniques. These challenges can impact the accuracy, consistency, and reliability of the decision-making process. Normalization in MADM faces challenges due to diverse data types, heterogeneous sources (varying scales, units, and formats), and sensitivity to technique. Nonlinear relationships complicate normalization processes, while the lack of a universal framework leads to subjectivity. Extreme values of decision criteria may distort results. The missing data requires careful handling to obtain appropriate normalization. Interpretability and practicality trade-offs arise, especially with complex normalization methods. Computational demands grow with large or dynamic datasets that have a higher number of decision criteria, where frequent updates necessitate renormalization. Finally, metric selection introduces bias, as no single approach suits all contexts. Addressing these issues requires adaptive, context-aware normalization to ensure reliable and consistent decision-making. This diversity in data and techniques may complicate the normalization process. These challenges highlight the complexity of selecting and applying the right normalization technique in MADM problems. Addressing these issues requires careful consideration of the problem context, data types, decision objectives, and available tools to ensure that the chosen normalization method enhances the overall decision-making process rather than introducing bias or errors.

MATLAB Code

Here is the MATLAB implementation code for all the normalization method which we have discussed in the normalization section. This section has MATLAB code that performs normalization for a considering a decision matrix X, with two sets of attribute indices: AT1 (benefit) and AT2 (cost):

```
%%%%%%%%%%%%%%%% Linear Normalization %%%%%%%%%%%%%%
  DM=[1000 40 13 10 31;1815 51 29 14 44; 2103 99 62 5 16;1624 76 47 17 39]; %Decision
  matrix
  AT1 = [1, 2, 3, 4]; % benefit criteria
  AT2 = [5];    % cost criteria
  [m, n] = size(DM); % m is the number of alternative and n is the number of criteria
  R = zeros(m, n); % normalized result matrix
  % Normalize benefit attributes
  for j = AT1
     max_val = max(DM(:, j));
     if max_val ~= 0
        R(:, j) = DM(:, j) / max_val;
     else
        R(:, j) = 0; % in case if max is zero
     end
  end
  % Normalize cost attributes
  for j = AT2
     min_val = min(DM(:, j));
     if min_val ~= 0
        R(:, j) = min_val ./ DM(:, j);
     else
        R(:, j) = 0;% in case if min is zero
     end
  end
  R
```

```
%%%%%%%%%%%%%%% Max Min Liner Normalization %%%%%%%%%%%%%%
  DM=[1000 40 13 10 31;1815 51 29 14 44; 2103 99 62 5 16;1624 76 47 17 39]; %Decision
matrix
  AT1 = [1, 2]; % benefit criteria
  AT2 = [3];    % cost criteria
  [m, n] = size(DM); % m is the number of alternative and n is the number of criteria
  R = zeros(m, n); % normalized result matrix
  xmin = min(DM); % Finding minimum of each column of decision matrix
  xmax = max(DM); % Finding maximum of each column of decision matrix
  range = xmax - xmin;
  % Normalize benefit criteria
  R(:, AT1) = (DM(:, AT1) - xmin(AT1)) ./ range(AT1);
```

```
% Normalize cost criteria
R(:, AT2) = (xmax(AT2) - DM(:, AT2)) ./ range(AT2)
R

%%%%%%%%%%%%%%%% Sum linear Normalization %%%%%%%%%%%%%%%%
DM=[1000 40 13 10 31;1815 51 29 14 44; 2103 99 62 5 16;1624 76 47 17 39]; %Decision
matrix
AT1 = [1, 2, 3, 4]; % benefit criteria
AT2 = [5];    % cost criteria
[m, n] = size(DM); % m is the number of alternative and n is the number of criteria
R = zeros(m, n); % normalized result matrix
for j = 1:n
  if ismember(j, AT1) % Benefit-type attribute
    A = sum(DM(:, j));
    R(:, j) = DM(:, j) / A;

  elseif ismember(j, AT2) % Cost-type attribute
    A = sum(1 ./ DM(:, j));
    R(:, j) = (1 ./ DM(:, j)) / A;

  else
    error(['Attribute type of column ' num2str(j) ' is not specified in AT1 or AT2.']);
  end
end
R

%%%%%%%%%%%%%%%% Vector Normalization %%%%%%%%%%%%%%%%
DM=[1000 40 13 10 31;1815 51 29 14 44; 2103 99 62 5 16;1624 76 47 17 39]; %Decision
matrix
AT1 = [1, 2, 3, 4]; % benefit criteria
AT2 = [5];    % cost criteria
[m, n] = size(DM); % m is the number of alternative and n is the number of criteria
R = zeros(m, n); % normalized result matrix
% Perform vector normalization
for j = 1:n
  A = sqrt(sum(DM(:,j).^2)); % Denominator for column j
  if ismember(j, AT1)
    R(:,j) = DM(:,j) / A;
  elseif ismember(j, AT2)
    R(:,j) = 1 - (DM(:,j) / A);
  end
end
R
```

```
%%%%%%%%%%%% Enhanced Accuracy Normalization %%%%%%%%%%%
DM=[1000 40 13 10 31;1815 51 29 14 44; 2103 99 62 5 16;1624 76 47 17 39]; %Decision
matrix
  AT1 = [1,2, 3, 4]; % benefit criteria
  AT2 = [5];     % cost criteria
  [m, n] = size(DM); % m is the number of alternative and n is the number of criteria
  R = zeros(m, n); % normalized result matrix
% Loop through each criterion
for j = 1:m
  if ismember(j, AT1)
    xj_max = max(DM(:,j));
    A = sum(xj_max - DM(:,j));
    if A == 0
      R(:,j) = 1; % All values are equal, normalization gives 1
    else
      R(:,j) = 1 - (xj_max - DM(:,j)) / A;
    end
  elseif ismember(j, AT2)
    xj_min = min(DM(:,j));
    A = sum(DM(:,j) - xj_min);
    if A == 0
      R(:,j) = 1; % All values are equal
    else
      R(:,j) = 1 - (DM(:,j) - xj_min) / A;
    end
  end
end
R

%%%%%%%%%%%%%%% Logarithmic Normalization %%%%%%%%%%
DM=[1000 40 13 10 31;1815 51 29 14 44; 2103 99 62 5 16;1624 76 47 17 39]; %Decision
matrix
  AT1 = [1, 2, 3, 4]; % benefit criteria
  AT2 = [5];     % cost criteria
  [m, n] = size(DM); % m is the number of alternative and n is the number of criteria
  R = zeros(m, n); % normalized result matrix
  for j = 1:n
    % Compute the log of the product for column j
    log_product = log(prod(DM(:, j)));
    for i = 1:m
      log_xij = log(DM(i, j));

      if ismember(j, AT1)
      R(i, j) = 1 - (log_xij / log_product);
      elseif ismember(j, AT2)
        R(i, j) = (1 - (log_xij / log_product)) / (m - 1);
      else
        R(i, j) = NaN; % Or handle columns not in AT1 or AT2
      end
    end
  end
R
```

References

1. N. Vafaei, R.A. Ribeiro, L.M. Camarinha-Matos, Normalization techniques for multi-criteria decision making: analytical hierarchy process case study, in *Technological Innovation for Cyber-Physical Systems*, vol. 470, ed. by L.M. Camarinha-Matos, A.J. Falcão, N. Vafaei, S. Najdi (Springer International Publishing, Cham, 2016), pp. 261–269. https://doi.org/10.1007/978-3-319-31165-4_26
2. A.S. Milani, A. Shanian, R. Madoliat, J.A. Nemes, The effect of normalization norms in multiple attribute decision making models: a case study in gear material selection. Struct Multidisc Optim **29**(4), 312–318 (2005). https://doi.org/10.1007/s00158-004-0473-1
3. N. Vafaei, R.A. Ribeiro, L.M. Camarinha-Matos, Selection of normalization technique for weighted average multi-criteria decision making, in *Technological Innovation for Resilient Systems*, ed. by L. M. Camarinha-Matos, K. O. Adu-Kankam, M. Julashokri (Springer International Publishing, Cham, 2018), pp. 43–52. https://doi.org/10.1007/978-3-319-78574-5_4
4. A.K. Yadav, K. Singh, N.I. Arshad, M. Ferrara, A. Ahmadian, Y.I. Mesalam, MADM-based network selection and handover management in heterogeneous network: a comprehensive comparative analysis. Results Eng. **21**, 101918 (2024)
5. T.M. Lakshmi, V.P. Venkatesan, A comparison of various normalization in techniques for order performance by similarity to ideal solution (TOPSIS). IJCOA 3(3), 255–259 (2014). https://doi.org/10.20894/IJCOA.101.003.003.023
6. S. Zolfani, M. Yazdani, D. Pamucar, P. Zarate, A VIKOR and TOPSIS focused reanalysis of the MADM methods based on logarithmic normalization. arXiv preprint (2020). arXiv:2006.08150
7. S. Lee, Proven methods of log transformation for enhanced statistical accuracy, https://www.numberanalytics.com/blog/log-transformation-statistical-accuracy Accessed 14 Apr 2025
8. Mastering Log Transformation Techniques for Better Data Analysis, https://www.numberanalytics.com/blog/log-transformation-data-analysis. Accessed 14 Apr 2025
9. Standardization: Why Z-Scores are Essential in Statistics, https://www.statisticalaid.com/standardization/. Accessed 14 Apr 2025
10. What is: Z-Normalization—Understanding Standardization, https://statisticseasily.com/glossario/what-is-z-normalization-understanding-standardization/. Accessed 14 Apr 2025
11. Y.B. Gopisetty, H.R. Sama, An integrated MCDM approach using double normalization: introducing the DN-WENSLO and DN-RPEM methods for socio-economic performance evaluation. J. Oper. Res. Soc. Accessed 14 Apr 2025.https://doi.org/10.1080/01605682.2025.2486679
12. K. Nitta, Decision making, in *Problem solving, cognitive processes & strategies* (Britannica, 2025), https://www.britannica.com/topic/von-Neumann-Morgenstern-utility-function. Accessed 15 April 2025
13. N. Vafaei, R.A. Ribeiro, L.M. Camarinha-Matos, Assessing normalization techniques for simple additive weighting method. Procedia Computer Science **199**, 1229–1236 (2022). https://doi.org/10.1016/j.procs.2022.01.156
14. A.K. Yadav, K. Singh, 6—Impact of number and type of criteria on ranking abnormality in MCDM techniques, in *Uncertainty in Computational Intelligence-Based Decision Making*, ed. by A. Ahmadian, S. Salahshour, V. E. Balas, and D. Baleanu (Academic Press, 2025), pp. 77–86. https://doi.org/10.1016/B978-0-443-21475-2.00017-5

Chapter 3
Subjective Weight Method

Abstract This chapter discusses the necessity of distributing subjective criteria weights to each criterion of alternative when addressing complex decision problems. This chapter explores the core concept, principles, taxonomy, complexity, and MATLAB implementation associated with this approach. This chapter also explores the different benefits, appropriateness for diverse decision problems, and constraints associated with each model.

Keywords Decision support · Subjective weights · Expert judgment · Pairwise comparison · Analytic hierarchy process (AHP) · Fuzzy subjective weights · Criteria importance · Consistency ratio · Subjectivity in decision making

3.1 Need for Subjective Weight

There are three operations in utilizing any decision-making technique involving numerical analysis of alternatives: 1. Determine the relevant criteria and alternatives. 2. Attaching statistical measures to the relative importance of the criteria and the impacts of the alternatives on these criteria. 3. Ranking of each alternative. The need for the subjective weight method in decision problems arises from the recognition that not each decision criteria or factors hold equal significance in complex decision-making scenarios. Decision-makers leverage subjective weighting to assign varying degrees of significance to different criteria based on their judgment, expertise, or contextual relevance. This approach accommodates the inherently subjective nature of certain decision factors, allowing for a more nuanced and personalized approach to decision analysis. Subjective weight methods play a crucial role in addressing the variability in individual perspectives and priorities. Different stakeholders or decision-makers may perceive the importance of criteria differently, and subjective weighting provides a flexible mechanism to capture this diversity of viewpoints. In the realm of complex decision problems, not all criteria contribute equally to the overall objective or desired outcome. Subjective weight distribution enables a more nuanced evaluation, emphasizing crucial factors while downplaying

A. K. Yadav et al., *Multi-attribute Decision-Making: Bridging Theory and Practice*, Transactions on Computer Systems and Networks,
https://doi.org/10.1007/978-981-95-4782-1_3

less significant ones. This adaptability is particularly valuable when dealing with multifaceted decisions where a one-size-fits-all approach may be insufficient. Moreover, subjective weight methods enable decision-makers to account for intangible or qualitative factors that may not be easily quantifiable. In situations where certain criteria involve subjective judgments, emotions, or qualitative considerations, the flexibility of subjective weighting provides a means to integrate these elements into the decision-making process.

The need for subjective weight methods further extends to addressing uncertainty and ambiguity in decision-making. In complex scenarios where some criteria lack precise quantitative measures or have uncertain impacts, subjective weighting allows decision-makers to express their level of confidence or uncertainty in assigning weights. In circumstances where the decision-making process may be incomplete and uncertain, this approach may prove to be helpful in tackling the situation. The subjective weight approach promotes a more inclusive decision-making process by involving stakeholders in the weighting process. Engaging relevant parties in determining the importance of criteria fosters a sense of ownership and transparency, enhancing the overall acceptability of decision outcomes. This participatory approach not only considers diverse perspectives but also contributes to improved stakeholder engagement and consensus-building. In dynamic environments where the significance of criteria may evolve over time, subjective weighting allows for adaptability. Decision-makers can adjust weights based on changing circumstances, ensuring that the decision model remains relevant and responsive to evolving conditions. This adaptive quality makes subjective weight methods invaluable in decision problems characterized by dynamic and evolving factors. Subjective weight methods provide a means to capture the context-specific relevance of criteria, particularly in decision contexts where the importance of criteria may vary. This adaptability is particularly valuable in sectors or industries where contextual factors significantly influence decision outcomes. In situations where traditional models may not fully capture the complexity of a decision problem, subjective weights can introduce novel perspectives and unconventional criteria, fostering creative problem-solving and encouraging a more holistic exploration of potential solutions. Figure 3.1 shows the classification of weighting methods. In this chapter, we will only discuss the subjective method. The other two, objective and hybrid weighting methods, will be discussed in Chaps. 4 and 5, respectively.

The weights assigned to criteria in multi-criteria evaluation incorporate both qualitative and quantitative data to ensure more accurate and well-informed decision-making. However, assigning weights to each decision criterion based on qualitative data can be influenced by the level of knowledge, expertise, experience, and preferences of the decision-maker. To address this issue, Saaty [1] introduced a numerical scale from 1 to 9, where '1' represents equal importance and '9' indicates extreme importance, thereby transforming qualitative judgments into quantitative values [1]. Weight classification can also be grouped into three categories: subjective, objective, and integrated (or combined) weighting approaches [2]. Subjective weight determination relies on experts' opinions, level of knowledge, and experience. To gather these subjective judgments, analysts typically prepare and present

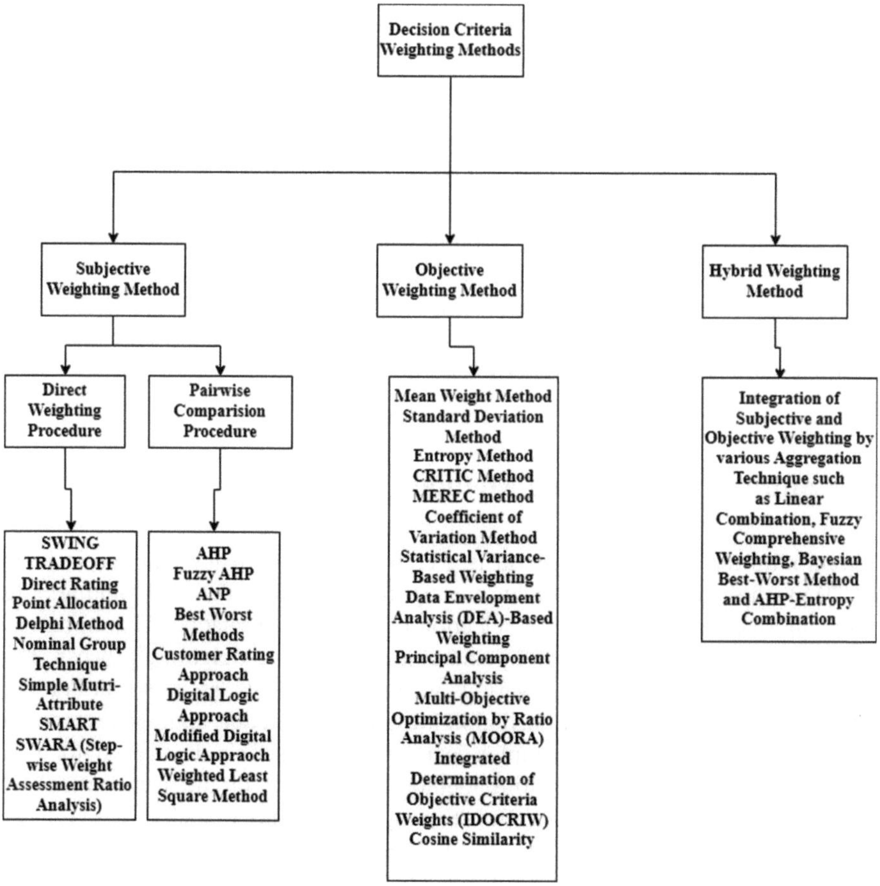

Fig. 3.1 Classification of weighting methods

structured questionnaires to decision-makers during the evaluation process. However, determining weights subjectively can be a time-consuming and complex task, especially when multiple decision-makers are involved and their opinions differ significantly. Reaching consensus often requires several rounds of discussions and revisions, which can delay the decision-making process and affect the consistency of the final results. Example of the subjective weighting method is the AHP, Olson [3] wrote on the subjectivity in multiple decision criteria decision analysis; he argued that judgment is at the heart of human decision-making and, therefore, considered judgment to be subjective. If a decision is to be made objectively, it is critical to adhere to the decision support perspective, which holds that the final judgment should rest with human decision-makers while acknowledging that all models have limitations. Models cannot account for all influencing factors, and even the most precise objective measurements will always have some degree of error or uncertainty. The author also emphasized that we should trust our own judgment as the ultimate deciding factor. In objective weighting methods, the weights for each criterion are calculated purely from the available data using mathematical models, without involving the decision-maker's personal input or preferences [4]. The integrated weighting approach combines both subjective and objective weighting methods to balance expert judgment with data-driven analysis. This hybrid method aims to overcome the limitations of using either approach alone by merging the strengths of both perspectives. Subjective weights are typically based on the experience, intuition, and preferences of decision-makers, while objective weights rely on quantitative data. Figure 3.1 illustrates some of the most commonly used subjective weighting techniques applied in MCDM processes.

3.2 Subjective Weighting Methods

Determining the weights of criteria is a common and critical challenge in many MCDM techniques, as these weights can significantly affect the final ranking order of the alternative and thereby the selection of the best alternative. Therefore, it is essential to consider the subjectivity and objectivity of the weighting process. This chapter provides an overview of various weighting methods, as shown in Fig. 3.1, used in multi-criteria optimization, highlighting several valuable approaches reported in the literature for addressing complex decision problems. This chapter will focus on some important subjective weighting methods, which rely on the decision-maker's knowledge, experience, and preferences to assign importance to each criterion. These methods are typically simpler and more straightforward in computation compared to objective weighting methods, which determine weights mathematically without input from the decision-maker. Subjective methods help ensure desirable decision-making properties and facilitate the evaluation of multiple performance measures to identify the most suitable alternative. Objective weighting techniques will be explored in detail in Chap. 4. The most common criteria weight method is the subjective method because of its simplicity and low complexity. Following are some vital subjective weighting methods:

1. Point allocation
2. Direct rating
3. Ranking method
4. Pairwise comparison (AHP)
5. Ratio method
6. Swing method
7. Delphi method
8. Nominal group technique
9. Simple multi-attribute ranking technique (SMART).

The further subsections will have brief discussions on some of the vital subjective weighting methods given in the above list.

3.2.1 The Point Allocation Method

This is one of the simplest methods used to determine criteria weights. According to the priority of criteria, a decision-maker allocates a certain number of points to each criterion. The more points a criterion receives, the greater its relative importance. This approach requires the decision-maker to distribute 100 points among the criteria they are evaluating. The total of all criterion weights must sum up to 100. This method is easy to normalize. However, the weights obtained from the point allocation method are not very precise, and the method becomes more difficult as the number of criteria increases to six or more. The point allocation method is a straightforward approach to determine the weights of criteria based on their priority. In this method, a decision-maker distributes a fixed number of points among the criteria, indicating their relative importance. While this method is easy to normalize, making comparisons more straightforward, it has limitations. The weights derived from the point allocation method are not highly precise, and the process becomes more challenging when dealing with a larger number of criteria, especially six or more. The point allocation method is a subjective weighting method used in MADM. This method involves decision-makers allocating a certain number of points to each attribute to express their preferences or priorities. The total number of points assigned to all attributes is usually fixed, helping decision-makers emphasize the importance of certain attributes over others. The following are the general steps involved in the point allocation method.

1. Define Point Scale: Decision-makers first establish a point scale that represents the relative importance or priority of each decision criteria. This scale typically ranges from 0 to a maximum value, with the total points across all attributes adding up to a constant value (e.g., 100 points).
2. Allocate Points: Decision-makers then distribute points among the attributes based on their perceived importance. Higher point allocations indicate a higher level of importance or preference for a specific attribute.
3. Normalization: To ensure consistency and comparability across decision-makers, or when comparing different decision scenarios, it may be necessary to normalize the points. Normalization involves adjusting the point allocations to create a weighted percentage representation.
4. Analysis: Once the points are allocated and normalized, the decision-makers can analyze the results to understand the relative importance of each attribute. This analysis can guide the decision-making process by highlighting the attributes that carry more weight in the overall decision.

5. Sensitivity Analysis: Sensitivity analysis may be conducted to assess how changes in the points assigned to each attribute affect the overall decision. This helps decision-makers understand the robustness of their preferences.

The point allocation method is straightforward and allows decision-makers to express their preferences in the form of a numerical value and in a structured manner. However, like any subjective weighting method, it is subject to the biases and perspectives of the decision-makers involved. It is important to consider the context of the decision and ensure that the point allocations accurately reflect the decision-makers' priorities. While the method provides a clear and transparent way to communicate preferences, it is crucial to be aware of the potential limitations, such as the assumption that decision-makers can accurately quantify and express the relative importance of different attributes in a numerical form.

3.2.2 Swing Method

The Swing weighting method is a multi-criteria analysis tool used to determine criteria weights through structured comparisons made by decision-makers. It involves evaluating and ranking features relative to a benchmark deemed the most important. Both the SWING method and its other variants, such as P- Swing, provide a systematic approach to facilitate clear understanding and effective application of weighting techniques. These methods are part of a broader set of MCDM approaches, including SMART, swing weighting, and the AHP. Each method comes with its strengths and limitations, influenced by factors such as resource demands and the potential for decision-making bias. The Swing weighting method is a subjective weighting method commonly used in MADM. It allows decision-makers to express their preferences by adjusting the weights assigned to different decision criteria based on their relative significance. This method is particularly useful when decision-makers have a general sense of the importance of attributes but may find it challenging to assign specific numerical values. Following is the general steps involved in a in how the Swing weighting method works:

1. Initial Weight Assignment: Decision-makers start by assigning initial weights to the attributes based on their subjective judgment.
2. Sensitivity Analysis: The decision-makers then conduct a sensitivity analysis to examine the impact of variations in the weights on the overall decision.
3. Adjustment Process (Swinging): Decision-makers "swing" the weights back and forth, adjusting them iteratively until they reach a satisfactory balance. This involves redistributing weights among attributes to reflect changing priorities.
4. Consensus Building: If multiple decision-makers are involved, the method may include a consensus-building phase where the participants discuss and negotiate the weights until an agreement is reached.
5. Final Weight Assignment: The final set of weights reflects the collective judgment or consensus of the decision-makers.

In the swing weighting method, the decision-maker is asked to select an alternative with the worst outcome and picks the criteria whose performance is likely to change (or swing) from its worst to the best. The criterion with the most preferred swing is given higher weight in terms of point value. Next, the criteria whose performance the decision-maker would like to change from its worst to the best level is selected again, and a value between 0 and the highest representing its relative importance regarding the most important criteria is provided. Then obtain the average normalized weights and the normalized weight interval [5]. The consistency rule for the swing weight method is crucial and necessary to ensure consistency of the weights assigned to individual cells of the matrix. In assigning swing weights, the stakeholders need to make compromises between the level of importance and the level of variation in the measure scale. This is done by allowing the stakeholders to assign an arbitrarily large weight to the top left-hand side of the matrix. The following expression can be used to calculate the normalized swing value for the ith value.

$$w_i = \frac{C_i}{\sum_{i=1}^{n} C_i}$$

where C_i is the unnormalized swing weight. The swing weighting method is flexible and allows decision-makers to respond to changes in preferences or priorities during the decision-making process. It acknowledges that the initial weight assignments may not be fixed and can be adjusted based on ongoing discussions and reflections. It is important to note that while the Swing method is a useful tool, it relies heavily on the subjective input of decision-makers. As a result, it is essential to be aware of potential biases and ensure that the method is applied transparently. Sensitivity analysis can help evaluate the robustness of the final decision with respect to changes in the weights. Overall, the swing method provides a structured approach for decision-makers to iteratively refine their preferences and reach a consensus on the relative importance of different attributes in a multi-attribute decision-making context [6].

3.2.3 Direct Rating Method

The direct rating method is one of the subjective weighting methods of MADM that have an approach where the decision-maker initially ranks all decision criteria based on their significance. In this method, decision-makers assign subjective ratings or scores to each criterion based on their relative significance in the decision-making process. The rating does not constrain the decision-maker's responses as the fixed-point scoring methods does. They can modify the importance of one criterion without needing to adjust the weight of another [7, 8]. The direct rating method does not impose constraints on the decision-maker's responses. Following are general steps involved in the direct rating method:

1. Ranking Criteria: Decision-makers evaluate and rank all criteria based on their perceived importance or relevance to the decision at hand.
2. Rating Scale: Decision-makers assign subjective ratings or scores to each criterion. The rating scale is often numerical, and decision-makers may use a scale (e.g., 1 to 10) to express the relative importance of each criterion.
3. Flexibility: Unlike some other weighting methods, the direct rating method allows decision-makers the flexibility to adjust the importance of one criterion without affecting the weights of others. This flexibility can be valuable in reflecting nuanced preferences.
4. Analysis: The ratings assigned to each criterion are then used to calculate the overall importance or weight of each criterion in the decision-making process.

The direct rating method is appreciated for its simplicity and ease of implementation. However, it does rely heavily on the subjective judgment of decision-makers, and the lack of a standardized scale may introduce variability in interpretations. Since it does not explicitly involve a fixed-point system, decision-makers need to be careful to maintain consistency in their rating scales. Despite its limitations, the direct rating method can be a practical tool when decision-makers have clear insights into the relative importance of criteria and desire a straightforward approach to express their preferences in a decision-making context.

3.2.4 Pairwise Comparisons

Pairwise comparisons involve the analysis of multiple populations or criteria in pairs to assess their significant differences. In decision-making, this method requires the decision-maker to compare each criterion with others and ascertain their preference levels for each pair. An ordinal scale (typically ranging from 1 to 9) is employed to quantify the preference value of one criterion compared to another. One well-known method that utilizes pairwise comparisons is the Analytical Hierarchy Process (AHP). This section has details discussion on AHP weighting method.

T. L. Satty, an operation research expert, developed AHP in the 1970s [9]. It is a well-known subjective method for distributing weights for criteria in MADM problems. The AHP breaks down difficult decision-making issues into top-down hierarchies, in which the top level is functionally independent of all of the sublevels, and attributes within each level are also functionally independent of one another. In general, a complicated problem can typically be broken down into three levels in the AHP process: the top level, the second level, and the bottom level, which can be goals, criteria, and alternatives, respectively. In AHP, complex problems are broken down into more manageable subproblems, which are then solved by the formulation of a pairwise comparison matrix. The processes of hierarchical structuring and prioritization enable these comparison matrix analyses to provide both flexibility and consistency checks. On the other hand, AHP needs all network criteria to be independent, and the cost of computing pairwise comparisons increase as the number of criteria used goes up. It is used to compute the weight of criteria of alternatives. The

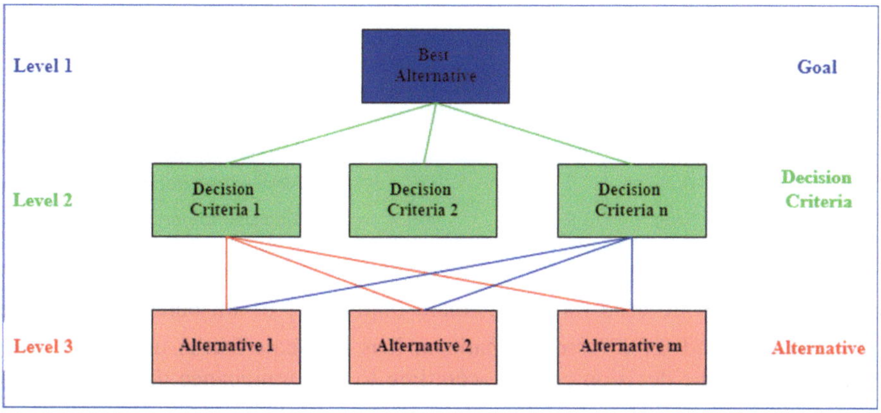

Fig. 3.2 AHP hierarchical structure based on goal, criteria and alternative

first and foremost step in AHP is creating a hierarchical structure shown in Fig. 3.2 in which the goal (network selection in our problem) is kept at the first level. The pairwise comparison matrix gives the relationship between variables in terms of the importance of criteria to criteria with respect to the goal on the basis of the Saaty nine-point scale. The size of the pairwise comparison matrix is determined by the number of criteria that are used to evaluate alternatives. The values of the pairwise comparison matrix depend on various decision-makers and are determined by the entity that wants to rank the alternatives. For finding the value decision-maker as the question "how much important criteria i is over criteria j to achieve the goal by alternative k?" The consistency ratio is used to test the correctness of assigned weight to criteria.

A pairwise matrix make use of a predefined value in the range [1–9] that is determined by the relative significance of the characteristics of each alternative. The decision criteria makes a comparison between the values of each attribute in order to determine the final priority weight of each attribute for each of the available alternatives [10–12]. The results of the pairwise comparison of each attribute have shown in a square matrix $A = \left[a_{ij}\right]_{n\times n}$. This matrix is also referred to as a pairwise comparison matrix.

$$A = \begin{bmatrix} a_{11} & a_{12} \cdots & a_{1n} \\ a_{21} & a_{22} & a_{2n} \\ \vdots & \ddots & \vdots \\ a_{n1} & a_{n2} \cdots & a_{nn} \end{bmatrix}$$

where n is the number of decision criteria of each alternative. The number of available alternatives is represented by m. The value of a_{ij} represents the ith decision criteria with respect to j^{th} decision criteria. Matrix A satisfy the condition $a_{ij} \neq 0$, $a_{ij} = 1/a_{ij}$and $a_{ij} = 1$.

Table 3.1 Defined importance scale for decision factors [13, 14]

Importance weight	Attribute importance
1	Equal importance
3	Moderate importance
5	Strong importance
7	Very strong importance
9	Extreme importance

2, 4, 6, and 8 are used for intermediary importance of attributes and $^1/_3$, $^1/_5$, $^1/_7$, and $^1/_9$ used for inverse comparison. It is the scale that is utilized in order to characterize the decision criteria according to the significance that they hold in relation to QoS, and it is expressed as a weight of criteria. The next step in the AHP process to figure out how important each of the alternatives' features is. AHP procedure has the following step to find out the weight distribution of decision criteria of alternative.

Step 1. Develop a hierarchical structure with a common goal at the first, criteria at 2nd, and alternative at 3rd level.

Step 2. Compare the relative importance through Table 3.1 of each decision criterion to the common goal using a pairwise comparison matrix. The value of diagonal elements is always one in the obtained pairwise comparison matrix.

Step 3. Pairwise comparison matrix can be normalized using any suitable normalization method. For example, the following vector normalization formula can be used to convert a pairwise comparison matrix into its normalized form.

$$r_{ij} = \frac{a_{ij}}{\sum_{i=1}^{n} a_{ij}}$$

Step 4. For computing the subjective criteria weight ($W_{\text{AHP}i}$) of the alternatives use the following formula

$$W_{\text{AHP}i} = \frac{\sum_{j=1}^{n} r_{ij}}{n}$$

For verification and consistency of the weight distribution, the concept of consistency ratio is used. The consistency ratio (CR) is ratio between consistency index (CI) and random index (RI), i.e., $CR = CI/RI$ where $CI = \frac{\lambda_{\max}-n}{n-1}$, Where $\lambda_{\max}$ is maximum eigenvalue from above matrix. Where RI is random index of a pairwise matrix generated at random. It is detailed in the following Table 3.2 where n is the total number of assumed criteria in decision-making problem

If $CR < 0.1$, then pairwise matrix process has a better consistency. Otherwise, we need to adjust the pairwise comparison matrix until the judgment matrix process satisfies the consistency condition. A smaller CI means that the judgment matrix is

Table 3.2 Random index scale [14]

n	1	2	3	4	5	6	7	8	9	10
RI	0.00	0.00	0.58	0.90	1.12	1.24	1.32	1.41	1.45	1.45

close to complete consistency. AHP method of weight computation considers each attribute independently. It is particularly useful when quantifying the decision results directly and accurately [15–17].

Some complex decision-making problem may have sub criteria in decision-making which can affect the second level that ultimately affect the overall decision-making process. These sub criteria put at third level and alternative at fourth level. It is possible that there may be some disagreement as to why subcritical cannot be taken at the second level. If we keep the sub-criteria at the second level, it becomes more difficult to compare them. It is simple to compare all of the sub-criteria that fall under a single criterion. Each alternative has a value of sub criteria of a criteria. To calculate weight of attribute in case of alternative having sub criteria the first and foremost step to develop pairwise comparison matrix between criteria of level two in hierarchy and find weights associated with them. Then individually compare sub criteria which fall under single criteria i.e., we need to compare sub criteria 1,2, 3…n under criteria 1 and so on. Finally calculate the weight associated with each sub criteria. So, there will be local weight and global weights of each sub criteria and criteria. To find the global weights of criteria we required to multiply the weights of criteria in second level to weights of sub criteria at level three. Finally formulate DM and solve it using the global weights of criteria or sub criteria [16, 18].

3.2.4.1 Advantage of AHP

1. It incorporates pairwise comparisons for each decision criterion, making judgments more accurate than when assessing all alternatives at once. This method reduces cognitive overload and ensures decisions are more reliable. Moreover, it also has the concept of consistency checks within the comparisons, helping decision-makers identify and correct any logical inconsistencies, leading to more robust outcomes.
2. The decision-maker is not required to give precise numerical values; rather, verbal, or qualitative judgments are sufficient for the evaluation. This approach reduces the burden on the decision-maker and makes the process more intuitive, while still providing meaningful input for analysis and comparison of alternatives.
3. It allows for a hierarchical organization of criteria, helping decision-makers concentrate more effectively on specific criteria and subcriteria when assigning weights.
4. It employs a ratio scale, which means comparisons can be made without needing specific units of measurement.

5. It enables the evaluation of both qualitative and quantitative criteria and alternatives using a unified preference scale.

3.2.4.2 Disadvantage

1. The pairwise comparison matrix may contain potential inconsistencies, which can lead to misleading or unreliable results. If the decision-maker's judgments are not logically consistent, the calculated priorities may not accurately reflect true preferences, ultimately affecting the validity and credibility of the final decision or ranking.
2. In case the number of criteria or alternatives is high, the demanding pairwise comparisons may increase the complexity of the problem and decrease the consistency of pairwise comparisons.
3. As the number of alternatives increases, the number of indirect comparisons also grows, requiring more extensive calculations and resulting in longer processing times.

3.2.5 *Fuzzy AHP*

In the traditional AHP, the scale of relative importance is defined using fixed crisp numerical values, where each linguistic term corresponds to a specific number. For example, the term moderate is assigned a value of 3. However, this rigid assignment can be problematic, as it fails to capture the inherent uncertainty and subjectivity involved in human judgment. For example, how do we assign a numerical value to a preference that falls between moderate (3) and strong (5)? Should it be 3.5? Should we interpret it as moderate or strong? To address these limitations, the fuzzy AHP approach was developed. In fuzzy AHP, linguistic terms are represented using fuzzy numbers rather than crisp values. A fuzzification process achieves this goal by transforming qualitative judgments into membership functions that represent the degree of belonging to each linguistic category. Through this method, precise numbers such as 1 to 9 are replaced by fuzzy scales that better accommodate the ambiguity and imprecision of human decision-making [19, 20]. Table 3.3 shows a fuzzy scale of relative significance of linguistic variables that is used in fuzzy AHP methodology.

Table 3.3 Predefined importance scale for comparing decision criteria in Fuzzy AHP [21]

Importance weight	Attribute importance
1, 1, 1	Equal importance
2, 3, 4	Moderate importance
4, 5, 6	Strong importance
6, 7, 8	Very strong importance
9, 9, 9	Extreme importance

The remaining fuzzy numbers (1, 2, 3), (3, 4, 5), (5, 6, 7), and (7, 8, 9) are used to represent intermediate levels of importance between the primary linguistic terms. The fuzzy AHP process first define the fuzzy pairwise comparison matrix, where each element is represented by a fuzzy membership number to capture the uncertainty in judgments. Once the matrix is established, the next step is to compute the fuzzy synthetic extent value S for each criterion. This value represents the aggregated preference of a criterion relative to all others. After obtaining the synthetic extents, the degree of possibility $V\left(S_i > S_j\right)$ is calculated to compare the significance of one criterion over another. For each decision criterion, the minimum degree of possibility across all comparisons is then determined, representing the least favorable condition under which the criterion can still be considered important. Finally, these minimum degrees are normalized to produce the final weights, reflecting the relative priority of each criterion in the decision-making process. The Fuzzy AHP weighting method, as proposed by Buckley (1985), involves the following steps to compute the weights of the criteria [22, 23].

Step 1: Construct a fuzzy pairwise comparison matrix and compute its reciprocal (A^{-1}) using the following function. Here, each criterion is represented by a fuzzy number in the form (l, m, u), where l, m, and u denote the lower, middle, and upper values, respectively.

$$A^{-1} = (l, m, u)^{-1} = \left(\frac{1}{u}, \frac{1}{m}, \frac{1}{l}\right)^{-1}$$

Step 2: Compute fuzzy geometric value $(\tilde{r}_i)$ by considering (l_i, m_i, u_i) and $\left(l_j, m_j, u_j\right)$ two fuzzy number of criteria.

$$\tilde{A}_i \otimes \tilde{A}_j = (l_i, m_i, u_i) \otimes \left(l_j, m_j, u_j\right) = \left(l_i \times l_j, m_i \times m_j, u_i \times u_j\right)$$

$$\tilde{r}_i = (l_1 \times l_2 \times \ldots l_n, m_1 \times m_2 \ldots \times m_n, u_1 \times u_2 \ldots \times u_n)^{1/n}$$

Step 3: Fuzzy geometric weight (w_i)

$$w_i = \tilde{r}_i \otimes (\tilde{r}_1 \oplus \tilde{r}_2 \ldots \oplus \tilde{r}_n)^{-1}$$

$$(r_1 \oplus r_2 \ldots \oplus n) = (l_1 + l_2 + \ldots + l_n, m_1 + m_2 + \ldots + m_n, u_1 + u_2 + \ldots + u_n)$$

Step 4: Defuzzify the fuzzy weights (w_i) using centroid method to get the crisp numerical values.

$$w_i = \left(\frac{l + m + u}{3}\right)$$

In case of total weight greater than one which is not acceptable, weight must be normalized so sum of all criteria weight become one using $w_i/\sum w_i$ [24].

Fuzzy AHP is an extension of the traditional AHP designed to handle the vagueness and uncertainty in human judgments by using fuzzy numbers instead of crisp values. Fuzzy AHP weighting method is particularly more useful when decision-makers are uncertain about their preferences or when the decision environment is vague. Fuzzy AHP has the capability to handle uncertainty and imprecision in human judgments. It has the potential to capture the imprecision of human judgments, especially when experts are unsure or hesitant. Unlike classic AHP, it does not force decision-makers to assign exact numerical values. It allows the use of linguistic scales such as moderate, strong that are converted into fuzzy numbers, making it more intuitive. The use of triangular fuzzy numbers leads to more realistic and flexible weight computation, improving result reliability. Fuzzy AHP is an extension of AHP, so it retains AHP's hierarchical structure while enhancing its ability to deal with ambiguity. It effectively aggregates judgments from multiple experts by accommodating variability in opinions through fuzzy aggregation. Fuzzy AHP is an alternative that can better model real-world decision-making scenarios. This implementation provides a basic framework that can be extended with more sophisticated fuzzy operations or additional defuzzification methods as needed for specific applications.

Irrespective of the above advantage, on the other hand, it also has some disadvantages. Compared to standard AHP, fuzzy AHP requires more computational complexity, including fuzzy arithmetic and comparisons, which can be more complex. Selecting appropriate fuzzy numbers for linguistic terms is subjective and may vary between practitioners or domains. Fuzzy logic concepts like membership functions, synthetic extent, and degree of possibility may not be easily understood by non-experts, which may limit its understandability. As the number of decision criteria or alternatives increases, the number of fuzzy pairwise comparisons grow rapidly, making it time. There is no universally accepted method for fuzzification or defuzzification, which can lead to inconsistencies in results across studies [25].

3.2.6 Comparison of Conventional AHP and Fuzzy AHP

Following are the key differences between AHP and Fuzzy AHP based on various features [26–28].

This comparison shown in Table 3.4 indicates that fuzzy AHP is an extension of conventional AHP that incorporates fuzzy set theory to better handle real-world decision-making scenarios where judgments are often imprecise or uncertain [26, 29].

Table 3.4 Comparison between AHP (analytic hierarchy process) and Fuzzy AHP

Characteristics	Conventional AHP	Fuzzy AHP
Basic concept	Uses crisp numbers for pairwise comparisons	Uses fuzzy numbers (e.g., triangular fuzzy numbers) for pairwise comparisons
Type of input	Crisp values (exact numbers: 1–9)	Fuzzy numbers (typically triangular fuzzy numbers: (l, m, u))
Weight calculation	Eigenvector method or geometric mean	Extent analysis, logarithmic least squares, or other fuzzy methods
Handling of uncertainty	Does not handle uncertainty or vagueness in judgments	Designed to handle ambiguity and vagueness using fuzzy logic
Judgment scale	Fixed, numerical scale (Saaty's scale)	Linguistic terms converted into fuzzy numbers
Complexity	Simpler and computationally efficient	More complex due to fuzzy arithmetic and comparisons
Accuracy of human judgment	Less accurate in uncertain environments	More accurate representation of subjective and imprecise judgments
Interpretability	Easy to understand and implement	Requires understanding of fuzzy set theory and membership functions
Aggregation of judgments	Direct averaging of crisp values	Aggregation of fuzzy numbers, better suited for group decisions
Consistency check	Uses consistency ratio	Consistency is not explicitly checked in most fuzzy AHP variants
Use of membership functions	Not used	Essential for representing fuzzy judgments
Applications	Suitable for problems with well-defined and precise criteria	Preferred when judgments are subjective, vague, or imprecise

3.2.7 Ranking Method

This is among the most straightforward ways to determine the weights for decision criteria in evaluating alternatives. In this method, the criteria are typically ordered based on their relative importance, starting from the most important down to the least significant. Once the ranking is established, the weights can be calculated using one of three ranking techniques. These are the Rank Sum Method, where weights are based on the sum of ranks; the Rank Exponent Method, which uses an exponential formula to emphasize rank differences; and the Rank Reciprocal Method, where weights are derived from the reciprocal of each rank, giving higher ranks greater influence. Following are concise descriptions of each method:

3.2.7.1 Rank Sum Method

It is simple and quick, best when subjective ranking is reliable. This method assigns weights based on the order of ranks, assuming equal step differences. This method computes the weights from the individual ranks, normalizing them by dividing the sum of the ranks [30]. The formula for rank sum determination can be expressed as

$$w_j^{\mathrm{RS}} = \frac{n - P_j + 1}{\sum_{k=1}^{n} n - P_k + 1}$$

where P_j is the rank of the jth criterion, $j = 1, 2\ldots, n$.

3.2.7.2 Rank Exponent Method

The rank exponent method is similar to the rank sum method but introduces an exponent to adjust the influence of each rank. In this approach, each rank value is raised to the power of a parameter p, which is usually determined by the decision-maker to reflect the emphasis placed on the most important criterion. This exponent allows for greater flexibility in differentiating criteria importance. The general formula for the rank exponent method is shown below [30].

$$w_j^{\mathrm{RE}} = \frac{\left(n - P_j + 1\right)^P}{\sum_{k=1}^{n} \left(n - P_j + 1\right)^P}$$

where P_j is the rank of the jth criterion, and P is the parameter describing the weights, $j = 1, 2, \ldots, n$.

3.2.7.3 Rank Reciprocal Method

It is also known as inverse weights method. The reciprocal weights method uses the normalized reciprocal of the criterion rank. This can be expressed as [30].

$$w_j^{\mathrm{RR}} = \frac{{}^{1}/_{P_j}}{\sum_{k=1}^{n} \left({}^{1}/_{P_k}\right)}$$

where P_j is the rank of the jth criterion, $j = 1, 2\ldots, n$.

These methods are generally unsuitable when dealing with a large number of decision criteria because directly ranking numerous criteria becomes challenging and often impractical. Nevertheless, they are still valuable as simple weight approximation techniques. Their main advantage lies in their ease of use, providing a quick and straightforward way to estimate and verify the relative importance of different criteria. As such, they serve as convenient tools for an initial assessment before applying more complex weighting methods.

3.2.8 Nominal Group Technique (NGT)

The Nominal Group Technique is a structured method for brainstorming that encourages every member of a group to contribute ideas equally on a given topic or issue. It not only helps generate a wide range of ideas but also provides a systematic way to rank and prioritize them. Typically, each participant writes down ideas individually, which are then shared and discussed with the group. After discussion, participants privately rank or vote on the ideas. The ideas with the highest number of votes or ranks are considered the most important or feasible aspects and are selected for further action or implementation [31, 32]. In NGT the weights are derived by carrying out the following steps with experience people in a group of not less than seven members.

Step 1: Individual Idea Generation on Paper: In this step, each participant is given approximately 10 min to independently and silently generate as many ideas as possible related to the issue at hand.

Step 2: Round-Robin Recording of Ideas on a Flip Pad: In this step, each participant is asked to share the best idea from their list generated during the first round. All ideas are recorded on a flip pad or whiteboard, and this process continues in a round-robin fashion until every idea from each participant's list has been presented.

Step 3: Discussion of Ideas for Clarification: In this step, all the ideas are discussed for better clarification. The facilitator goes through the master list from the beginning, asking participants whether each idea is clear. If any idea is unclear for any reason, it must be clarified by the person who suggested it or by someone else in the group.

Step 4: Voting to Select the Most Important Ideas: This final step is crucial because the selected ideas will be used to determine the criteria weights. In this step, each participant is asked to choose the five most important ideas from the master list and rate them on a scale from 1 to 5 based on their importance. The most important idea receives a rating of 5, while the least important receives a rating of 1. Every participant provides their ratings individually, and the aggregated scores are then calculated accordingly.

NGT gathers individual expert judgments on the importance of various criteria, combines them systematically, and uses group discussion and ranking to reach a consensus. It is especially useful in subjective or qualitative decision-making environments. The NGT provides several advantages. It is a structured and systematic method that encourages equal participation from all group members, ensuring that each expert's opinion is considered independently before any group discussion begins. This helps in minimizing the risk of groupthink or dominance bias, where stronger personalities might otherwise influence the outcome. Because NGT emphasizes consensus-building through a transparent and auditable process, it enhances the credibility and acceptance of the final weight values among stakeholders. Furthermore, NGT is especially useful in qualitative decision environments, where expert judgment is crucial due to the lack of measurable data [31, 33].

Irrespective of the above advantage, on the other hand, NGT also has some disadvantages and limitations. The process can be time-consuming and resource-intensive, especially when dealing with a large number of criteria or a large panel of experts. Its limited scalability makes it less practical for very complex MCDM problems. Moreover, while it is structured, the resulting weights are still subjective, relying heavily on human judgment, which may lead to inconsistencies across different expert groups. The technique also requires a skilled and neutral facilitator to guide the process effectively, as any bias in facilitation can affect the quality of the outcomes. Finally, the method of aggregating individual rankings into a single set of consensus weights may lead to compromises that obscure minority but potentially important opinions [32, 34].

3.2.9 Simple Multi-Attribute Rating Technique (SMART)

The SMART technique is a compensatory multiple criteria decision-making method originally developed by Edward in 1971 [35]. The SMART method rates alternatives and weights decision criteria to facilitate decision-making. This approach focuses on how to assign weights to criteria based on their relative importance. Specifically, the decision-maker is asked to rank each decision criterion from the least to the most important. This ranking helps ensure that more significant criteria have greater influence on the final decision. By following this structured process, the SMART method simplifies complex multi-criteria evaluations. This method assigns 10 points to the least important criterion and 100 points to the most important. The other criteria are given points in between, reflecting their relative importance. This approach is based on the principle that each alternative is evaluated against multiple criteria, each weighted according to its significance. The SMART method follows a structured sequence of steps, beginning with determining the number of criteria to be considered in the decision-making process. Next, each decision criterion is assigned a value on a scale from 1 to 100, giving higher scores to the most important decision criteria. These values are then normalized by comparing them to the total points assigned across all criteria to ensure consistency in weighting. Once the weights are set, parameter values for each criterion are specified for every alternative being evaluated. To calculate the utility value, each decision parameter value is converted into a standardized format, allowing fair comparison across decision criteria. The final score for each criterion is determined by normalizing the standardized value with its corresponding weight and summing these results for each alternative. This process ensures that criteria with higher importance have a greater influence on the overall decision. Finally, the weight of each criterion is obtained by dividing its assigned points by the total points so that all weights sum to one, providing a clear and balanced evaluation framework. The steps involved in the SMART method are as follows:

1. Specify the number of criteria involved in the evaluation.
2. Assign a value between 1 and 100 to each decision criterion, giving higher values to those with greater importance.
3. Normalize each criterion by dividing its assigned value by the sum of all criteria values to obtain the criteria weights
4. Assign parameter values for each criterion for every alternative
5. Calculate the utility value by converting the parameter values of each criterion into standardized data values.
6. Determine the final value for each criterion by multiplying the normalized standard data value by the normalized weight of the criterion, and then summing these products.

The SMART method is a multi-criteria decision-making approach used to evaluate alternatives based on several criteria. Each alternative is assessed by assigning values to its criteria, while each criterion is given a weight to indicate its relative importance compared to others. This combination of values and weights helps in ranking and selecting the best possible alternative objectively. A key advantage of the SMART method is the simplicity of the questions posed to the decision-maker and the ease with which the responses can be analyzed. In the SMART method, ratings of alternatives are assigned directly, in the natural scales of the criteria. In a view to keeping the weighting of the criteria and the rating of the alternatives needs to be separate as possible; the different scales of criteria need to convert into a standard internal scale. In this technique, the decision-maker performs the process mathematically by applying a value function. The simplest and most widely used form of a value function method is the additive model, which in the easiest cases can be applied using a linear scale.

3.2.9.1 Pros and Cons of SMART Method

The straightforward nature of the questions and answers is a major advantage of the SMART method. This simplicity directly enhances the decision-maker's understanding of the process used to solve the problem. Unity refers to a linear function and represents the simplest form of a value function, commonly used due to its straightforward applicability in most cases involving linearity. This model assumes interdependence, meaning it remains unaffected by the specific set of alternatives considered. It incorporates all lowest-level attributes, ensuring comprehensive evaluation in the decision-making process. Additionally, the model emphasizes relevancy by maintaining consistent decision scores regardless of the number of alternatives, as the ratings are not significantly relative. This stability makes the linear value function a practical and reliable choice for many decision-making scenarios where simplicity and consistency are prioritized.

On other hand, as the number of criteria increases, the method becomes more complex, making it challenging to manage and interpret. While it can improve accuracy by eliminating poorly performing alternatives during the initial screening phase, it may not adequately address variations in scale ranges when determining appropriate category weights, which can affect the sensitivity of the results. Moreover, the technique often lacks consistency due to its inherently subjective nature, as judgments and preferences can vary among decision-makers, potentially leading to inconsistent outcomes across different evaluations.

SMART is a widely adopted method known for its simplicity and ease of use. It is applicable across various fields such as urban planning, business, engineering, and environmental studies. When addressing complex MCDM problems, SMART proves to be an effective approach. This technique is particularly efficient when used by a single decision-maker rather than a group. However, SMART should be viewed as a decision support tool rather than a definitive solution. Since identifying the optimal solution manually can be challenging, SMART provides a practical and approximate indication of the most suitable outcome [36].

The decision-makers may have personal biases, preferences, or cognitive limitations that influence their judgment, which leads to inconsistent or irrational weight assignments to decision criteria. Because different decision-makers may assign different weights to the same problem, results may not be reproducible, reducing reliability and credibility. The core principle of subjective weighting methods depends on the requirement of domain experts who may not always be available or may disagree, which can limit its scalability and applicability, especially in data-rich environments. In subjective methods, even minor changes in opinions can drastically alter the weight structure, potentially impacting decision outcomes and causing instability, as well as increasing sensitivity to subjective views. Subjective methods, such as the AHP, involve pairwise comparisons that increase exponentially with the number of criteria. Large-scale or real-time applications find it unfeasible. Subjective methods are not designed for automated, data-driven environments, which makes them not suitable for large volumes of data effectively. From the above discussion, we have observed that subjective weighting methods in MCDM rely on human knowledge, expertise, experience, and judgment to assign weights to criteria. Subjective weighting methods, such as AHP and others, are widely used but come with several limitations and challenges that have forced the development of objective weighting approaches. In Chaps. 4 and 5, there are detailed discussions on the objective and hybrid/comprehensive weighting methods, respectively.

3.3 Matlab

3.3.1 MATLAB Code for AHP

Following is the MATLAB implementation of the Burkley method for AHP, a widely accepted approach to Analytic Hierarchy Process. This implementation can be customized to suit the specific requirements of any given problem.

```
% Step 1: Define the pairwise comparison matrix
% Example: 3 criteria — Cost, Quality, Time
A = [1   3   0.5;
   1/3  1   1/7;
   2    7   1];
n = size(A,1);
% Step 2: Normalize the matrix
column_sum = sum(A);
A_normalized = A ./ column_sum;
% Step 3: Calculate priority vector (weights)
weights = mean(A_normalized, 2);
% Step 4: Consistency Check
% Weighted sum vector
weighted_sum = A * weights;
% Consistency vector
consistency_vector = weighted_sum ./ weights;
% Lambda max
lambda_max = mean(consistency_vector);
% Consistency Index (CI)
CI = (lambda_max - n) / (n - 1);
% Random Index (RI) values for 1 to 10 items
RI_values = [0, 0, 0.58, 0.90, 1.12, 1.24, 1.32, 1.41, 1.45, 1.49];
RI = RI_values(n);
% Consistency Ratio (CR)
CR = CI / RI;
% Display results
disp('--- AHP Results ---');
disp('Normalized Pairwise Matrix:');
disp(A_normalized);
disp('Priority Vector (Weights):');
disp(weights);
disp(['Lambda max: ', num2str(lambda_max)]);
disp(['Consistency Index (CI): ', num2str(CI)]);
disp(['Consistency Ratio (CR): ', num2str(CR)]);
if CR < 0.1
  disp('The comparison matrix is consistent.');
else
  disp('The comparison matrix is NOT consistent. Please revise your judgments.');
end
```

3.3.2 MATLAB Code Fuzzy APH

The following is the MATLAB implementation of Fuzzy AHP, which utilizes Triangular Fuzzy Numbers and the Extent Analysis Method developed by Chang (1996), a widely recognized approach for Fuzzy AHP. It can be modified as per the specific requirement of any problem.

```
% Number of criteria
n = 3;
% Define Triangular Fuzzy Numbers (L, M, U) for each pairwise comparison
% Example matrix for 3 criteria:
%        C1      C2      C3
% C1  [1 1 1]  [2 3 4]  [4 5 6]
% C2  [1/4 1/3 1/2]  [1 1 1]  [3 4 5]
% C3  [1/6 1/5 1/4]  [1/5 1/4 1/3]  [1 1 1]
% Define the fuzzy comparison matrix (n x n x 3)
A(:,:,1) = [1   2   4;   % L values
           1/4 1   3;
           1/6 1/5 1];

A(:,:,2) = [1   3   5;   % M values
           1/3 1   4;
           1/5 1/4 1];
A(:,:,3) = [1   4   6;   % U values
           1/2 1   5;
           1/4 1/3 1];
% Step 1: Sum fuzzy values for each column
sumL = sum(A(:,:,1));
sumM = sum(A(:,:,2));
sumU = sum(A(:,:,3));
% Step 2: Normalize each element by total column sum
for i = 1:n
   for j = 1:n
      N(i,j,1) = A(i,j,1)/sumU(j); % L / U
      N(i,j,2) = A(i,j,2)/sumM(j); % M / M
      N(i,j,3) = A(i,j,3)/sumL(j); % U / L
   end
end
% Step 3: Sum across rows to get S_i = sum_row(N_ij)
for i = 1:n
   S(i,1) = sum(N(i,:,1));
   S(i,2) = sum(N(i,:,2));
   S(i,3) = sum(N(i,:,3));
end
% Step 4: Compute degree of possibility V(Si ? Sj)
V = zeros(n);
for i = 1:n
   for j = 1:n
      if i ~= j
         if S(i,2) >= S(j,2)
            V(i,j) = 1;
         elseif S(j,1) >= S(i,3)
            V(i,j) = 0;
         else
            V(i,j) = (S(i,3) - S(j,1)) / ((S(i,3) - S(i,2)) + (S(j,2) - S(j,1)));
         end
      end
```

```
    end
end
% Step 5: Minimum degree of possibility for each criterion
for i = 1:n
    d(i) = min(V(i, [1:i-1, i+1:end]));
end
% Step 6: Normalize to get weights
weights = d / sum(d);
% Display results
disp('Fuzzy AHP Weights:');
disp(weights');
```

3.4 Summary

This chapter explores subjective weighting methods, their purpose, types, and key techniques, along with step-by-step procedures, advantages, disadvantages, and associated challenges. It includes MATLAB implementations of AHP and fuzzy AHP. The chapter concludes with a discussion on the driving forces behind the evolution of objective weighting methods, highlighting the need for balance between expert judgment and data-driven approaches in multi-criteria decision-making processes.

References

1. D. Pamučar, Ž. Stević, S. Sremac, A new model for determining weight coefficients of criteria in mcdm models: full consistency method (fucom). Symmetry **10**(9), 393 (2018)
2. R. Ginevičius, V. Podvezko, Objective and subjective approaches to determining the criterion weight in multicriteria models (2005), https://etalpykla.vilniustech.lt/handle/123456789/138753. Accessed 13 Jan 2025
3. L.D. Olson, Subjectivity in multiple criteria decision analysis. Hum. Cent.Ed Process. **23**(4), 433–448 (2008)
4. A. Aldian, M.A. Taylor, A consistent method to determine flexible criteria weights for multicriteria transport project evaluation in developing countries. J. East. Asia Soc. Transp. Stud. **6**, 3948–3963 (2005)
5. G. Parnell, T. Trainor, 2.3.1 using the swing weight matrix to weight multiple objectives. INCOSE Int. Symp. **19**, 283–298 (2009). https://doi.org/10.1002/j.2334-5837.2009.tb00949.x
6. J. Mustajoki, R.P. Hämäläinen, A. Salo, Decision support by interval SMART/SWING—incorporating imprecision in the SMART and SWING methods. Decis. Sci. (2005). Accessed 21 May 2025. https://doi.org/10.1111/j.1540-5414.2005.00075.x
7. A. Arbel, Approximate articulation of preference and priority derivation. Eur. J. Oper. Res. **43**(3), 317–326 (1989)
8. M. Kress, Approximate articulation of preference and priority derivation—a comment. Eur. J. Oper. Res. **52**(3), 382–383 (1991)
9. P.T. Harker, L.G. Vargas, The theory of ratio scale estimation: saaty's analytic hierarchy process. Manage. Sci. **33**(11), 1383–1403 (1987). https://doi.org/10.1287/mnsc.33.11.1383
10. T.L. Saaty, Decision making with the analytic hierarchy process. Int. J. Serv. Sci. **1**(1), 83–98 (2008)
11. L.G. Vargas, An overview of the analytic hierarchy process and its applications. Eur. J. Oper. Res. **48**(1), 2–8 (1990). https://doi.org/10.1016/0377-2217(90)90056-H

12. T.L. Saaty, What is the analytic hierarchy process? in *Mathematical models for decision support* (Springer, 1988), pp. 109–121
13. T.L. Saaty, L.G. Vargas, Estimating technological coefficients by the analytic hierarchy process. Socioecon. Plann. Sci. **13**(6), 333–336 (1979). https://doi.org/10.1016/0038-0121(79)90015-6
14. R.W. Saaty, The analytic hierarchy process—what it is and how it is used. Math. Model. **9**(3–5), 161–176 (1987)
15. F. Zahedi, The analytic hierarchy process—a survey of the method and its applications. Interfaces **16**(4), 96–108 (1986). https://doi.org/10.1287/inte.16.4.96
16. X. Zeshui, W. Cuiping, A consistency improving method in the analytic hierarchy process. Eur. J. Oper. Res. **116**(2), 443–449 (1999)
17. A. Ishizaka, A. Labib, Analytic hierarchy process and expert choice: benefits and limitations. OR Insight **22**(4), 201–220 (2009). https://doi.org/10.1057/ori.2009.10
18. A. Ishizaka, A. Labib, Review of the main developments in the analytic hierarchy process. Expert Syst. Appl. **38**(11), 14336–14345 (2011)
19. J.J. Buckley, Ranking alternatives using fuzzy numbers. Fuzzy Sets Syst. **15**(1), 21–31 (1985)
20. Y. Liu, C.M. Eckert, C. Earl, A review of fuzzy AHP methods for decision-making with subjective judgements. Expert Syst. Appl. **161**, 113738 (2020)
21. C.-C. Sun, A performance evaluation model by integrating fuzzy AHP and fuzzy TOPSIS methods. Expert Syst. Appl. **37**(12), 7745–7754 (2010)
22. J.J. Buckley, Fuzzy hierarchical analysis. Fuzzy Sets Syst. **17**(3), 233–247 (1985)
23. D.-Y. Chang, Applications of the extent analysis method on fuzzy AHP. Eur. J. Oper. Res. **95**(3), 649–655 (1996)
24. L.C. Leung, D. Cao, On consistency and ranking of alternatives in fuzzy AHP. Eur. J. Oper. Res. **124**(1), 102–113 (2000)
25. K. Mukherjee, A note on limitations of FAHP. Springer. Accessed 22 May 2025. https://doi.org/10.1007/978-81-322-3700-6_5
26. A. Özdağoğlu, Comparison of AHP and fuzzy AHP for the multi- criteria decision making processes with linguistic evaluations
27. A Ishizaka, Comparison of fuzzy logic, AHP, FAHP and hybrid fuzzy AHP for new supplier selection and its performance analysis. Int. J. Integr. Supply Manag. Accessed 22 May 2025. https://doi.org/10.1504/IJISM.2014.064353
28. R. Mosadeghi, J. Warnken, R. Tomlinson, H. Mirfenderesk, Comparison of Fuzzy-AHP and AHP in a spatial multi-criteria decision making model for urban land-use planning. Comput. Environ. Urban Syst. **49**, 54–65 (2015). https://doi.org/10.1016/j.compenvurbsys.2014.10.001
29. G. Kabir, Comparative analysis of AHP and fuzzy AHP models for multicriteria inventory classification. Int. J. Fuzzy Log. Syst. (2011)
30. E. Roszkowska, Rank ordering criteria weighting methods–a comparative overview. Optim. Stud. Ekon. **5**(65), 14–33 (2013)
31. M.M.B. Abdullah, R. Islam, Nominal group technique and its applications in managing quality in higher education. Pak. J. Commer. Soc. Sci. **5**(1), 81–99 (2011)
32. S.S. McMillan, M. King, M.P. Tully, How to use the nominal group and Delphi techniques. Int. J. Clin. Pharm. (2016). https://doi.org/10.1007/s11096-016-0257-x
33. N. Harvey, C.A. Holmes, Nominal group technique: an effective method for obtaining group consensus. Int. J. Nurs. Pract. Accessed 22 May 2025. https://doi.org/10.1111/j.1440-172X.2012.02017.x
34. W.M. Fox, The improved nominal group technique (INGT). J. Manag. Develop. **8**(1), 20–27 (1989). https://doi.org/10.1108/EUM0000000001331
35. M. R. Patel, M. P. Vashi, B. V. Bhatt, *SMART—Multi-criteria decision-making technique for use in planning activities* (2017)
36. G.O. Odu, Weighting methods for multi-criteria decision making technique. J. Appl. Sci. Environ. Manag. **23**(8), 1449 (2019). https://doi.org/10.4314/jasem.v23i8.7

Chapter 4
Objective Weight Methods

Abstract This chapter discusses the necessity of distributing objective criteria weights to each criterion of alternative to indicate significance of each decision criteria during the addressing complex decision problems. This chapter explores the core concepts, principles, complexity, MATLAB implementation, and research gaps associated with various weight estimation approaches. This chapter also explores the different benefits, appropriateness for diverse decision problems, and constraints associated with each model.

Keywords Objective weights · Data-driven weights · Contrast intensity · Correlation coefficient · Standard deviation · Conflict intensity · Entropy · CRITIC · MEREC · SVP

4.1 Objective Weight

In MADM, objective weights involve assigning quantitative values to each decision criteria based on measurable and reproducible factors. The objective weights, derived from data analysis, enable intelligent decisions by reducing subjective biases. These weights provide a systematic and transparent representation of each criterion's importance. Objective weights enhance decision-making by reducing ambiguity, allowing for comparability among alternatives, and supporting mathematical models like weighted sum and outranking methods. This method of appropriate weight distribution proves instrumental in risk management, resource prioritization, and sensitivity analysis-based decision-making problems. Although objective weights are derived from data, there may still be a degree of subjectivity, especially when defining criteria [1].

A. K. Yadav et al., *Multi-attribute Decision-Making: Bridging Theory and Practice*, Transactions on Computer Systems and Networks,
https://doi.org/10.1007/978-981-95-4782-1_4

4.2 Need for Objective Weight

Subjective weights are determined based on individual knowledge, judgment, beliefs, experiences, or preferences, influenced by personal opinions, and subject to variation among individuals. In contrast, objective weights are established through measurable and quantifiable criteria grounded in factual data, calculations, or explicit criteria to minimize subjectivity. Subjective weights, being reliant on personal knowledge and experience and lacking a clear quantitative basis, are challenging to measure objectively. Conversely, objective weights are quantifiable and often represented by numerical values derived through data analysis. Subjective weights may lack transparency as they are influenced by individual factors that may not be clearly expressed or comprehensible to others. On the other hand, objective weights are generally more transparent and easily communicable, rooted in explicit criteria or quantitative measures that can be clearly explained and documented. Subjective weights are prone to variations and inconsistencies across different decision-makers or situations due to individual biases and perspectives. But objective weights tend to be more consistent, relying on clear and measurable criteria that are less susceptible to individual biases. Subjective weights are commonly used in situations where qualitative considerations, personal values, or subjective decisions play a substantial role in decision-making. In comparison, objective weights are preferred in scenarios where decision-makers seek a more data-driven, systematic, and transparent approach, particularly when dealing with criteria that can be objectively measured [2].

Achieving reliable, robust, and efficient MADM methods requires careful consideration of both objective and subjective weight in estimating the stable ranking of alternatives in complex decision problems. The comprehensive weight estimation through the integration of subjective and objective weights is discussed in the next chapter. The objective weights in MADM provide a systematic, transparent, and accountable framework for aligning decision-making with data and improving the effectiveness of decision support systems. In situations where such conflicts arise, objective criteria computation methods like CRITIC, entropy, mean, and standard deviation prove to be most suitable. We will discuss various objective weight methods in further sections, along with their MATLAB implementation, advantages, and limitations [3].

4.3 Objective Weight Methods

Several objective methods exist, grounded in different principles. In this segment, we have explored some of the vital objective weight methods that may be crucially applied in various emerging applications contexts. The following weight method is included in our further discussion:

1. Entropy
2. CRiteria Importance Through Intercriteria Correlation (CRITIC)

3. Improved CRiteria Importance Through Intercriteria Correlation (I-CRITIC)
4. MEthod based on the Removal Effects of Criteria (MEREC)
5. Improved MEthod based on the Removal Effects of Criteria (I-MEREC)
6. Statistical Variance Procedure (SVP)
7. Standard Deviation
8. Mean.

4.3.1 Decision Problem Definition

For further discussion of various weighing methods to compute and assign weight to each decision criteria, let us assume that the decision matrix $DM = \left[x_{ij}\right]_{m\times n}$ can be used to define the multi-criteria decision problem, where $a_1, a_2, \ldots a_m$ are alternatives, and $c_1, c_2, \ldots c_n$ criteria of each alternatives. The entry in decision matrix x_{ij} represents values of ith alternative, with respect to jth criteria of DM. The decision criteria used in a decision problem may fall into two categories: cost (ψ) or benefit ($\emptyset$) criteria. The definition of a decision problem provided above will be applied to the methods explored in this section.

4.3.2 Entropy Method

The entropy method is an objective approach for allocating criteria weights to each decision criteria, particularly beneficial when decision-makers hold conflicting perspectives on the importance of criteria in decision problems. In contrast to methods like subjective weight, where decision-makers express opinions in pairwise comparisons based on their knowledge, expertise, and experiences, the entropy method operates without such subjective input. The entropy weight measures the similarity of various alternatives based on specific criteria and facilitates the comparison of two or more alternatives. Originating from a transportation model employing entropy to measure trip dispersion between origins and destinations, the weights computed through this method are termed objective weights [4, 5]. The following steps are involved in calculating weight employing entropy method:

Step 1: Compute the normalized decision matrix having element r_{ij}

$$r_{ij} = \begin{cases} \frac{x_{ij}}{\sum_{i=1}^{m} x_{ij}}, & if\ j \in \emptyset \\ \frac{\frac{1}{x_{ij}}}{\sum_{i=1}^{m} \frac{1}{x_{ij}}}, & if\ j \in \psi \end{cases}$$

Step 2: Compute entropy information e_j

$$e_j = -h \sum_{i=1}^{m} r_{ij} \ln r_{ij}$$

where $h = \frac{1}{\ln m}$

Step 3: Compute the weight vector w_j of each eriteria

$$w_j = \frac{1 - e_j}{\sum_{j=1}^{n} (1 - e_j)}$$

$(1 - e_{j)}$ is known as degree of diversity function. A low entropy value indicates a system with minimal disorder, corresponding to a high weight. In scenarios in which decision-makers want to address and minimize uncertainty in decision process, entropy method is particularly more appropriate.

4.3.3 CRITIC Method

The CRITIC approach of criteria weight estimation, introduced by chemical engineer D. Diakoulaki in 1995, is an effective method for estimating the objective weight of relative importance. This method takes into account the contrast in intensity and the inherent conflict within the framework of decision problems to determine the weights of criteria. CRITIC becomes particularly valuable when decision-makers hold conflicting views on weight values or face challenges in comparing different criteria [6, 7]. The computation of criteria weights for alternatives in the CRITIC approach has following steps:

Step 1: Obtain the normalized decision matrix having elements r_{ij} by normalizing the decision matrix using the following equation:

$$r_{ij} = \frac{x_{ij} - x_j^{\text{worst}}}{x_j^{\text{best}} - x_j^{\text{worst}}}$$

Step 2: Calculate the standard deviation (σ_j) for each criterion based on normalized values.

Step 3: Compute symmetric matrix of $n \times n$ with element r_{jk}, which is linear correlation coefficient between vector x_j and x_k. In case of same criteria, the linear correlation coefficient will be one. So diagonal value will be one.

Step 4: Compute the degree to which criterion j creates a conflict with the decision-making situation as defined by the other criterion i.e., $\sum_{k=1}^{n} (1 - r_{jk})$.

Step 5: Assess the quantity of information c_j pertinent to each of the criteria.

$$c_j = \sigma_j \times \sum_{k=1}^{n} (1 - r_{jk})$$

Step 6 Compute objective weight w_j

$$w_j = \frac{c_j}{\sum_{k=1}^{m} c_j}$$

This approach attributes significant weight to decision criteria exhibiting a high standard deviation and a weak correlation with other criteria. Moreover, the assessment procedure operates autonomously without requiring human involvement. Contrast strength gauges the information disparity between the same criterion across two different alternatives, while conflict evolution measures the shared information magnitude of a criterion belonging to both alternatives within the same alternative. These insights can be employed to analyze problems related to the selection of the best alternative network.

In assessing the significance of criteria through the CRITIC method, both the standard deviation and correlation coefficient are employed in calculations to ascertain contrast strength and conflict evolution, respectively. In the CRITIC method, conflict evolution is estimated by the correlation coefficient, which might have a correlation value of positive or negative between criteria i and j. Therefore, it might not be suitable to estimate a conflict using these formulas, given that the absolute values of a negative correlation coefficient and a positive correlation coefficient are identical.

4.3.4 *Improved CRITIC Method*

One of the most significant problems with the traditional CRITIC weighting method is that it is difficult to accurately reflect contrast strength when attempting to measure discrimination of various indicators using standard deviation. This is caused by the fact that different criteria use different measurement units and magnitude orders. In the CRITIC method, conflict evolution is computed by the correlation coefficient, which might be positive or negative between criteria i and j. So, it could be suitable to approximate a conflict using correlation formulas because the magnitudes of negative and positive correlation coefficients are the same.

Because of the robust mathematical concept and the objectivity inherent in the entropy approach, the improved CRITIC technique has replaced standard deviation with entropy information during calculating contrast strength to address the shortcomings of the CRITIC technique. The improved CRITIC technique has sufficient potential to objectively assess the consistency among the same criteria across various alternatives. It can objectively reflect the factors that are influential between the various heterogeneous alternatives. Additionally, it has the potential to mitigate the irrationality that has been brought on by the negative correlation coefficient that can be used when determining the conflict. The improved CRITIC has following steps to calculate weight of $\left[x_{ij}\right]_{m\times n}$:

Step1: The normalization of decision matrix can be achieved using the following equation where $\left[r_{ij}\right]_{m\times n}$ is normalized decision matrix.

$$r_{ij} = \begin{cases} \frac{x_{ij}}{\sum_{i=1}^{m} x_{ij}}, & if\ j \in \emptyset \\ \frac{\frac{1}{x_{ij}}}{\sum_{i=1}^{m} \frac{1}{x_{ij}}}, & if\ j \in \psi \end{cases}$$

Step 2: Calculate the entropy information (EI_j) for each criterion, diverging from the use of standard deviation.

$$EI_j = -\frac{1}{\ln(m)} \sum_{j=1}^{j=n} r_{ij} \ln(r_{ij})$$

Step 3: Create a symmetric matrix of $n \times n$ by calculating the linear correlation coefficient r_{jk} between criterion x_j and x_k for each element where $1 \le i, j \le n$.

$$\left[r_{jk}\right]_{n\times n} = \begin{cases} 1, & if\ j = k \\ \text{corr}\left(x_j, x_k\right), & if\ j \neq k \end{cases}$$

where $1 \le i, j \le n$

Step 4: At this stage, the measure of the conflict $M(c)$ that has been caused by x_j is computed using following equation, and it is compared to the decision situation that has been established by remaining criteria.

$$M(c) = \sum_{k=1}^{n} \left(1 - \left|r_{jk}\right|\right)$$

Step 5: Through the following equation, this step has provided an estimation of the quality of the information, denoted by QI_j, in relation to each of the criteria.

$$QI_j = EI_j \times \sum \left(1 - \left|r_{jk}\right|\right)$$

Step 6: The objective weight w_j of criterion.

$$w_j = \frac{QI_j}{\sum_{j=1}^{m} QI_j}$$

4.3.5 MEREC

This subsection introduces a recent criteria weight method called MEREC to determine criteria' objective weights. This method uses a novel idea for estimating the priority of each decision criterion. This method, based on the concept of the removal effects of criteria, is proposed to determine the criteria's weights in a MADM problem. The MEREC utilizes the impact of removing each criterion of alternative performance on alternative performance to assess criteria weights. Higher weights are assigned to criteria that have a more substantial impact on performances. This approach initially establishes a metric for evaluating alternative performances. It employs a straightforward logarithmic measure with uniform weights to compute the performance of alternatives. To assess the impact of removing each criterion, it employs the absolute deviation, which indicates the disparity between the overall performance of the alternative and its performance when a criterion is removed [8]. This method has utilized the following steps to compute the priority weight for each decision attribute for the alternative of the decision problem:

Step 1: The normalized decision matrix has been computed by utilizing the linear min–max normalization function. The computation of standardized decision has been done through following equation. Where elements (r_{ij}) of standardized decision matrix can be obtained as:

$$r_{ij} = \begin{cases} \frac{\min_j x_{ij}}{x_{ij}}, & if\ j \in \emptyset \\ \frac{x_{ij}}{\max_j x_{ij}}, & if\ j \in \psi \end{cases}$$

Step 2: Apply a nonlinear logarithmic function and weight each criterion equally to determine each alternative's performance (Z_i). The following equation gives the logarithmic function used to calculate alternative performance:

$$Z_i = \left(\log\left(1 + \frac{1}{n}\sum_i |\log(r_{ij})|\right)\right)$$

Step 3: Compute the performance (Z'_{kl}) after removing *jth* criteria by using following equation.

$$Z'_{ij} = \left(\log\left(1 + \frac{1}{n}\sum_{j \neq l} |\log(r_{ij})|\right)\right)$$

Step 4: Compute the actual absolute deviation of jth criteria (after removal of ith criteria) from ith criteria. After that calculate sum of all deviations. Let Ω_j denotes the sum of all absolute deviations and is calculated by following formula.

$$\Omega_l = \sum_i \left| Z'_{ij} - Z_i \right|$$

Step 5: Criteria weight w_j given by following equation.

$$w_j = \frac{\Omega_j}{\sum_j \Omega_j}$$

4.3.6 *Improved MEREC Method*

It is utilized in the process of calculating the criteria priority weight of decision criteria, which assists in making the decision regarding which alternative is best and most appropriate. The removal effect of each decision criteria on the performance of alternatives is used to figure out how the priority weight distribution of decision criteria should be distributed. This approach employs a logarithmic measure to compute alternative performance, assigning equal weights to each decision criterion. To assess the impact of removing each decision criterion, it utilizes the absolute deviation. The absolute deviation measure captures the difference between the performance when a decision criterion is removed and the overall performance [9]. This method has utilized following steps to compute to priority weight for each decision attributes for alternative of the decision problem.

Step 1: The computation of standardized decision has been done through following equation. Where elements (r_{ij}) of standardized decision matrix can be obtained as:

$$r_{ij} = \begin{cases} \frac{\min\limits_j x_{ij}}{x_{ij}}, & if\ j \in \emptyset \\ \frac{x_{ij}}{\max\limits_j x_{ij}}, & if\ j \in \psi \end{cases}$$

Step 2: The logarithmic function for calculating each alternative overall performance (Z_i) is provided in following equation.

$$Z_i = \sum_j \left|\log(r_{ij})\right| + \left(\log\left(1 + \frac{1}{n}\sum_j \left|\log(r_{ij})\right|\right)\right)$$

Step 3: Compute the performance (Z'_{kl}) after removing pth criteria by using following equation.

$$Z'_{kl} = \sum_{p \neq l} \left|\log(\eta_{kp})\right| + \left(\log\left(1 + \frac{1}{n}\sum_{p \neq l} \left|\log(\eta_{kp})\right|\right)\right)$$

Step 4: Compute the actual absolute deviation of *pth* criteria (after removal of *l*th criteria) from *l*th criteria. After that calculate sum of all deviations. Let Ω_l denotes the sum of all absolute deviations and is calculated by following formula.

$$\Omega_l = \sum_k \left|Z'_{kl} - Z_k\right|$$

Step 5: Criteria Weight w_j is given by following equation.

$$w_j = \frac{\Omega_j}{\sum_j \Omega_j}$$

4.3.7 SVP Method

The SVP method is an approach used in MADM to assign objective priority weights to each decision criteria. It works on the principle of variances, aiming to provide a systematic way to distribute priority weights to each decision criterion. In other words, SVP is a method for objectively allocating priority weights to criteria based on variances. The SVP method is employed to objectively prioritize criteria, especially when there is variability in the decision matrix. By considering the variances, it aims to highlight criteria with more consistent and stable performance in the decision-making process [10, 11]. The calculation of objectives using SVP comprises the following steps:

Step 1: Normalization enables to evaluate of opposing criteria within the same decision framework. The decision matrix normalizes using vectorially and transform into $\left[r_{ij}\right]_{m\times n}$ using following equation for beneficial and cost criteria:

$$r_{ij} = \begin{cases} \dfrac{x_{ij}}{\sqrt{(\sum_{i=1}^{n}\sum_{j=i}^{m}|x_{ij}|^2)}}, & if\ j \in \emptyset \\ \dfrac{x_{ij}}{\sqrt{\sum_{i=1}^{n}\sum_{j=i}^{m}|(\frac{1}{x}_{ij}|^2)}}, & if\ j \in \psi \end{cases}$$

Step 2: Calculating the weight of criteria using SVP

$$S_j = \sum_{i=1}^{m}\sum_{j=1}^{n}\frac{(r_{ij} - \overline{r}_{ij})^2}{m-1}$$

where $\overline{r}_{ij}$ represents the mean of the normalized values for each column in the normalized decision matrix.

Step 3: The ratio of S_j to the total variance of normalized decision matrix:

$$w_j = \frac{S_j}{\sum_{j=1}^{m} S_j}$$

where w_j is the objective weight of the *jth* criterion that the SVP assign. It is important to note that the effectiveness of the SVP method depends on the specific characteristics of the decision problem and the nature of the criteria involved. As with any MADM method, careful consideration of the decision context is crucial for meaningful and reliable results.

One of the positive aspects of SVP is that it allows for a quantitative evaluation of statistical variances within data sets. SVP allows for a quantitative evaluation of statistical variances within data sets, which is particularly effective in capturing and highlighting variations or dispersion within the data, offering insights into the degree of variability. But on the negative side, SVP relies on the assumption that the underlying data follows a normal distribution, which may not always be the case in real-world scenarios. Extreme values of the decision criteria can disproportionately impact variance calculations, potentially leading to skewed results and affecting the method's accuracy. The effectiveness of SVP may be influenced by the size of the sample, and caution should be exercised when applying it to small datasets. SVP is primarily designed for continuous data and may be less suitable for categorical or discrete data, limiting its applicability in certain situations [2].

4.3.8 Standard Deviation Method

The Standard deviation weight method is a technique used in MADM to assign weights to criteria based on the standard deviations of the normalized values of these criteria in the decision matrix. This method is particularly useful when decision-makers want to give higher importance to criteria that exhibit stability or consistency in their performance across alternatives. It aims to emphasize criteria with less variability, assuming that these criteria are more reliable or predictable in the decision-making process. Like any MADM method, the suitability of the standard deviation weight method depends on the specific characteristics of the decision problem [12]. The standard deviation weight method involves the following steps to estimate the assignment of importance to each decision criteria:

Step 1: Find the normalized decision matrix using the following equation:

$$r_{ij} = \frac{x_{ij}}{\sum_{i=1}^{m} x_{ij}}$$

Step 2: Find the standard deviation ($sd_{i)}$ for each criteria using following equation

$$sd_i = \sqrt{\frac{\sum_{i=1}^{m}\left(r_{ij} - \overline{r_j}\right)^2}{m}}$$

where $\overline{r_j}$ is the mean normalized value of criteria j, $1 \leq i \leq m$ and $1 \leq j \leq n$

Step 3: The weight of criteria

$$w_j = \frac{sd_i}{\sum_{j=1}^{n} sd_i}$$

4.3.9 Mean Weight Method

The use of the mean weight, which distribute equal significance to all the decision criteria, is typically favored in situations where there is a lack of input from the decision-maker or when insufficient information is accessible for deciding. This approach operates on the assumption that all criteria are of equal significance. The following equation can be used to derive weight information of each criterion where n is the number of criteria:

$$w_j = \frac{1}{n}$$

4.4 Comparative Analysis of Objective Weight Methods

The above listed objective weight methods differ in their approaches to assigning weights in decision-making processes. The Entropy method emphasizes the concept of information entropy from information theory. Primarily focuses on the level of uncertainty or disorder in the values of decision criteria. This method utilizes the information theory to measure the uncertainty or randomness in a set of decision criteria. In this method the weights are determined based on the degree of disorder or unpredictability. Entropy method is particularly more suitable when decision-makers want to address and minimize uncertainty in the decision process. CRIPTC method use the concept of pairwise correlations among criteria to determine their relative importance. The CRITIC estimates the criteria weigh based on contrast strength and correlation coefficient of attribute and alternatives. This method evaluates the inter-criteria correlations to establish the importance of each criterion. It focuses on relationships among criteria to derive their relative significance of each decision criteria weights. CRITIC is particularly useful when decision-makers are interested in understanding the interdependencies and relationships between different criteria. Improved CRITIC is an enhanced version of CRITIC, potentially incorporating refinements or adjustments to improve accuracy in assigning criterion weights. The improved version of CRITIC method used entropy information instead of standard definition which have the potential to resolve the issues of positive and negative correlation associated with CRITIC [13]. MEREC and Improved MEREC both method is utilizing the concept of the removal effects of criteria. It is one of the recent methods. SVP method utilizes statistical variance as a basis for determining the weights of decision criteria. In this the weights are influenced by the variability or spread of data associated with each criterion. The SVP method is particularly more relevant in situations where understanding the variability or spread of data is crucial in decision-making. The fundamental principle of standard deviation is measuring the dispersion of values around the mean, providing insight into the overall spread of data. This method determines weights based on the standard deviation of criteria, reflecting the degree of deviation from the mean. This method uses the concept of measures the dispersion of values around the mean. This method is more useful when decision-makers want to consider both the central tendency and the variability of criteria. However Mean method assigns equal importance to all criteria, if each criterion has the same weight. This method is more appropriate when decision-makers believe that each criterion should contribute equally to the decision [14].

So, from the above discussion it can be conclude that no weight method is perfect. The appropriability of the weighting method depend on nature of the decision problems. These objective weight methods employ diverse techniques, such as entropy, CRITIC, Improved CRITIC, MEREC, improved MEREC, statistical variance, and mean, to derive distribution of weights for decision criteria, reflecting varying considerations of relationships, uncertainty, and statistical properties. Each method has its strengths and may be suitable for different decision-making contexts. These methods vary in their theoretical foundations, considerations of uncertainty, treatment of inter-criteria relationships, and the balance between different statistical measures. The choice of method may depend on the specific requirements of the decision problem and the preferences of decision-makers.

4.5 Effectiveness of Various Weight Methods

Spearman's rank correlation coefficient is employed to assess the association between two sets of ranks generated by distinct methods. In this section, Spearman's rank correlation coefficient method has been used to evaluate and analyze the results of weighting methods. The presence of a high coefficient value indicates that there is a strong correlation between the two approaches [14]. The coefficient (r_s) is calculated through the following function:

Table 4.1 Decision matrix

Alternative/criteria	C1	C2	C3	C4	C5
A1	100	40	13	10	31
A2	1815	51	29	14	44
A3	2103	99	62	5	16
A4	1624	76	47	17	39

Table 4.2 Criteria weight obtained from methods

Methods/criteria	C1	C2	C3	C4	C5
Entropy	0.09017	0.16218	0.35644	0.22954	0.16168
CRITIC	0.21798	0.1501	0.1658	0.26431	0.2018
I_CRITIC	0.22993	0.14704	0.15813	0.26251	0.20238
MEREC	0.12128	0.2039	0.2848	0.22065	0.16938
I_MEREC	0.12393	0.21174	0.28317	0.21314	0.16802
SVP	0.09856	0.18053	0.32977	0.22695	0.16419
STD	0.20586	0.20118	0.19237	0.19848	0.20212

$$r_s = 1 - 6\frac{\sum d_i^2}{n(n^2 - 1)}$$

where d_i is the difference between the ranks of the two methods and n is the number of alternatives.

4.6 Numerical Computation and Analysis

For a quantitative analysis of the discussed method, we can utilize the decision matrix shown in Table 4.1, with all attributes treated as benefit criteria. Applying the mentioned methods to this matrix yields the weight of each decision criterion, as illustrated in Table 4.2 and Fig. 4.1. Subsequently, the criteria are ranked based on their weights, and Spearman's rank correlation coefficient can be employed to assess the relationship between the methods.

4.7 MATLAB Implementation

The implementation of the above-mentioned weight method may help researchers utilize it in their research work. So, this section has a discussion on the MATLAB implementation of the vital weight methods discussed above. Following is the

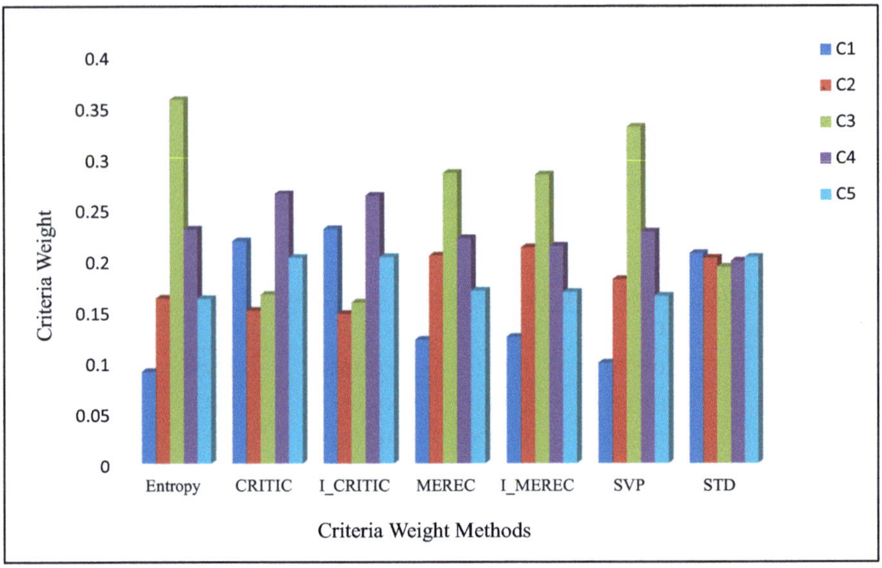

Fig. 4.1 Method versus alternative's criteria weight

MATLAB implementation of the entropy, CRITIC, Improved CRITIC, MEREC, Improved MEREC, SVP, and standard method.

4.7.1 Entropy Method

```
function [w_objective]=entropy(DM)
DM=[1000 40 13 10 31;1815 51 29 14 44; 2103 99 62 5 16;1624 76 47 17 39];
%DM=decision_matrix();
[m, n] = size(DM);
V = sum(DM);
featurewt = DM./repmat(V, m, 1);
k = 1/log(m);
sumfeature = sum(featurewt.*log(featurewt));
outputentropy = -k*sumfeature;
varcoeff = abs(1-outputentropy);
sumvarcoeff = sum(varcoeff,2);
w_objective = varcoeff/sumvarcoeff;
end
```

4.7.2 CRITIC

```
function [w_critic_weight_vector]=critic(DM)
%DM=decision_matrix();
DM=[1000 40 13 10 31;1815 51 29 14 44; 2103 99 62 5 16;1624 76 47 17 39];
[m,n]=size(DM);
Dmax=max(DM);
Dmin=min(DM);
par=0;
%MM vector for cost and benefit criteria
MM=[0 0 0 0 0];
if par==0
  for j=1:n
     if MM(j)==1
        PIS(j)= Dmax(j);
        NIS(j)= Dmin(j);
     else
        PIS(j)= Dmin(j);
        NIS(j)= Dmax(j);
     end
```

4.7.3 Improved CRITIC

```
function [w_improvedcritic_weight_vector]=improvedcritic(DM)
%DM=test1(1);
DM=[1000 40 13 10 31;1815 51 29 14 44; 2103 99 62 5 16;1624 76 47 17 39];
[m,n]=size(DM);
[alternatives, criteria] = size(DM);
summatrix = sum(DM);
featurewt = DM./repmat(summatrix, alternatives, 1);
k = 1/log(alternatives);
sumfeature = sum(featurewt.*log(featurewt));
outputentropy = -k*sumfeature;
St=outputentropy;
r=corr(featurewt);
r=abs(r);
E=ones(n,n);
C=St'.*sum((E-r),2);
sumC=sum(C);
w=C./sumC;
w_improvedcritic_weight_vector=w'
end
```

4.7.4 MEREC Method

```
function [w_MEREC_weight_vector]=critic(DM)
DM=[1000 40 13 10 31;1815 51 29 14 44; 2103 99 62 5 16;1624 76 47 17 39];
%DM=decision_matrix();
[m,n]=size(DM);
max1=max(DM);
min1=min(DM);
MM=[0 0 0 0 0];
for i=1:m
  for j=1:n
    if MM(j)==1
    V(i,j)=min1(j)/DM(i,j);
    elseif MM(j)==0
      V(i,j)=DM(i,j)/max1(j);
    end
  end
  end
  V;
  B=abs(log(V));
  C=sum(B');
  E=1+(1/4*C);
  F=log(E);
  V1=V(:,[2,3,4,5]);
  B1=abs(log(V1));
  C1=sum(B1');
  E1=1+(1/4*C1);
  F1=log(E1);
  G1=abs(F1-F);
  V2=V(:,[1,3,4,5]);
  B2=abs(log(V2));
  C2=sum(B2');
  E2=1+(1/4*C2);
  F2=log(E2);
  G2=abs(F2-F);
  V3=V(:,[1,2,4,5]);
  B3=abs(log(V3));
  C3=sum(B3');
  E3=1+(1/4*C3);
  F3=log(E3);
  G3=abs(F3-F);
  V4=V(:,[1,2,3,5]);
  B4=abs(log(V4));
  C4=sum(B4');
  E4=1+(1/4*C4);
  F4=log(E4);
  G4=abs(F4-F);
  V5=V(:,[1,2,3,4]);
  B5=abs(log(V5));
  C5=sum(B5');
  E5=1+(1/4*C5);
  F5=log(E5);
  G5=abs(F5-F);
  G=[F1' F2' F3' F4' F5'];
  H=[G1' G2' G3' G4' G5'];
  I=sum(H);
  w_MEREC_weight_vector=I./sum(I)
  end
```

4.7.5 Improved MEREC Method

```
function [w_I_MEREC_weight_vector]=critic(DM)
DM=[1000 40 13 10 31;1815 51 29 14 44; 2103 99 62 5 16;1624 76 47 17 39];
%DM=decision();
[m,n]=size(DM);
max1=max(DM);
min1=min(DM);
MM=[0 0 0 0 0];
for i=1:m
   for j=1:n
      if MM(j)==1
      V(i,j)=min1(j)/DM(i,j);
      elseif MM(j)==0
         V(i,j)=DM(i,j)/max1(j);
      end
   end
end
V;
B=abs(log(V));
C=sum(B');
D=log(C');
E=1+(1/n*C);
F=log(E)+D';
V1=V(:,[2,3,4,5]);
B1=abs(log(V1));
C1=sum(B1');
D1=log(C1');
E1=1+(1/n*C1);
F1=log(E1)+D1';
G1=abs(F1-F);
B2=abs(log(V2));
C2=sum(B2');
D2=log(C2');
E2=1+(1/n*C2);
F2=log(E2)+D2';
G2=abs(F2-F);
V3=V(:,[1,2,4,5]);
B3=abs(log(V3));
C3=sum(B3');
D3=log(C3');
E3=1+(1/n*C3);
F3=log(E3)+D3';
G3=abs(F3-F);
V4=V(:,[1,2,3,5]);
B4=abs(log(V4));
C4=sum(B4');
D4=log(C4');
E4=1+(1/n*C4);
F4=log(E4)+D4';
G4=abs(F4-F);
V5=V(:,[1,2,3,4]);
B5=abs(log(V5));
C5=sum(B5');
D5=log(C5');
E5=1+(1/n*C5);
F5=log(E5)+D5';
G5=abs(F5-F);
G=[F1' F2' F3' F4' F5'];
H=[G1' G2' G3' G4' G5'];
I=sum(H);
w_I_MEREC_weight_vector=I./sum(I)
end
```

4.7.6 SVP Method

```
function [w_SVP_weight_vector]=SVP(DM)
DM=[1000 40 13 10 31;1815 51 29 14 44; 2103 99 62 5 16;1624 76 47 17 39];
%DM=decision_matrix();
m,n]=size(DM);
tVec=sqrt(sum(DM.^2, 1));
V=DM./repmat(tVec,m,1);
t=mean(V);
x=(V-repmat(t,m,1)).^2;
y=sum(x);
e=y./(m-1);
w_SVP_weight_vector=e./sum(e);
end
```

4.7.7 Standard Deviation Method

```
function [w_std_weight_vector]=std(DM)
DM=[1000 40 13 10 31;1815 51 29 14 44; 2103 99 62 5 16;1624 76 47 17 39];
%DM=decision_matrix();
[m,n]=size(DM);
tVec=sqrt(sum(DM.^2, 1));
V=DM./repmat(tVec,m,1);
t=sum(V);
b=(V-repmat(t,m,1)).^2;
c=sum(b);
d=c./m;
e=sqrt(d);
f=sum(e);
w_std_weight_vector=e./f;
end
```

4.8 Summary

In this chapter, we have explored the concept of objective weight, delving into its significance, diverse methods for estimating criteria weight, conducting a comparative analysis, and examining both the advantages and limitations of these methods. Additionally, we have reviewed current research trends in weight estimation. The chapter includes a discussion of numerical examples, providing results from different weight estimation methods. Moreover, it has the MATLAB implementation of various approaches for estimating decision criteria weight.

References

1. A. Alinezhad, J. Khalili, *New methods and applications in multiple attribute decision making (MADM)*, vol. 277 (Springer, 2019)
2. R. Vavrek, Evaluation of the impact of selected weighting methods on the results of the TOPSIS technique. Int. J. Inf. Technol. Decis. Mak. **18**(06), 1821–1843 (2019)
3. C.-L. Hwang, A.S.M. Masud, *Multiple objective decision making—methods and applications: a state-of-the-art survey*, vol. 164 (Springer Science & Business Media, 2012), https://books.google.co.in/books?hl=en&lr=&id=M0noCAAAQBAJ&oi=fnd&pg=PA12&dq=Survey+of+objective+weight+methods+of+MADM&ots=0_QuTG3OEq&sig=9o2EEiKpZGQl85PPihpLv8iY42Q. Accessed 11 Dec 2023
4. I. Mukhametzyanov, Specific character of objective methods for determining weights of criteria in MCDM problems: entropy, CRITIC and SD. Decis. Mak.: Appl. Manag. Eng. **4**(2), 76–105 (2021)
5. Y. Zhu, D. Tian, F. Yan, Effectiveness of entropy weight method in decision-making. Math. Probl. Eng. **2020**, 1–5 (2020)
6. D. Diakoulaki, G. Mavrotas, L. Papayannakis, Determining objective weights in multiple criteria problems: the critic method. Comput. Oper. Res. **22**(7), 763–770 (1995)
7. S. Fujimoto, H. Hoof, D. Meger, Addressing function approximation error in actor-critic methods, in *Proceedings of the 35th International Conference on Machine Learning*, PMLR (2018), pp. 1587–1596, https://proceedings.mlr.press/v80/fujimoto18a.html. Accessed 11 Dec 2023
8. M. Keshavarz-Ghorabaee, M. Amiri, E.K. Zavadskas, Z. Turskis, J. Antucheviciene, Determination of objective weights using a new method based on the removal effects of criteria (MEREC). Symmetry **13**(4), 525 (2021)
9. A.K. Yadav, K. Singh, P.K. Srivastava, P.S. Pandey, I-MEREC-T: improved MEREC-TOPSIS scheme for optimal network selection in 5G heterogeneous network for IoT. Internet of Things, 100748 (2023)
10. A.A. Nasser, A.A. Alkhulaidi, M.N. Ali, M. Hankal, M. Al-olofe, A Weighted Euclidean distance-statistical variance procedure based approach for improving the healthcare decision making system in Yemen. Indian J. Sci. Technol. **12**(3), 1–15 (2019)
11. H.A. Tayali, M. Timor, Ranking with statistical variance procedure based analytic hierarchy process. Acta Infologica **1**(1), 31–38 (2017)
12. M.F. El-Santawy, R.A.Z. Eldin, SDV-VIKOR: a new approach for multi-criteria decision making with no preference. Life Sci. J. **4**, 10 (2013)
13. D. Liu, D. Zhang, J. Chen, Z. Pan, A heterogeneous network access selection algorithm based on improved CRITIC, in *2018 IEEE 3rd Advanced Information Technology, Electronic and Automation Control Conference (IAEAC)*. (IEEE, 2018), pp. 381–387
14. M. Şahin, A comprehensive analysis of weighting and multicriteria methods in the context of sustainable energy. Int. J. Environ. Sci. Technol. **18**(6), 1591–1616 (2021). https://doi.org/10.1007/s13762-020-02922-7

Chapter 5
Integrated Weight Methods

Abstract One of the critical challenges in decision-making is avoiding decisions becoming too subjective or too objective. To address this issue, the chapter focuses on discussion of integrating subjective and objective weighting methods, which provides a balanced approach to improve decision accuracy and reliability. This chapter explores the comprehensive weighting method, emphasizing its importance and outlining the different types available. It highlights the distinctions between comprehensive weighting and traditional subjective or objective approaches, examines the influence of proportion coefficients on weight assignment, and evaluates the effectiveness of various integrated methods. Additionally, the chapter covers the practical implementation of these methods using MATLAB and concludes with a summary of key insights.

Keywords Subjective · Objective · Comprehensive · Hybrid weight · AHP–Entropy · AHP–CRITIC · Proportion coefficients

5.1 Comprehensive Weight

MADM is an optimization technique that has the potential to address complex decision problems in science, engineering, and economics where diverse, complex, and conflicting decision criteria can be involved. Evaluating the significance of each decision criterion is critical in decision-making, as it significantly influences the consistency of the ranking order of alternatives. Evaluating the importance of each decision criterion is a critical step in the process of ranking and selecting optimal alternatives, especially when dealing with multiple conflicting criteria across different approaches. Researchers have grouped these weighting approaches to assigning weight to the criteria to each decision criteria into three categories: subjective, objective, and hybrid methods. This chapter focuses primarily on the comprehensive or hybrid weighting methods. The weight of the criteria for each decision criterion reflects the amount of information to assess the performance score of each alternative. For some complex decision problems, the subjective approach of assigning significance to each decision

A. K. Yadav et al., *Multi-attribute Decision-Making: Bridging Theory and Practice*, Transactions on Computer Systems and Networks,
https://doi.org/10.1007/978-981-95-4782-1_5

criterion is either difficult or unreliable due to the decision-maker's reliance on their knowledge, experience, and expertise. Therefore, the subjective weight method is not appropriate in some decision scenarios, as specific characteristics could exist independently of the decision-maker's knowledge, experience, expertise, perspective, and way of thinking. On the other hand, in the objective weighting method, decision criteria serve as the core component, functioning as decision-makers and sources of information. The objective weight method is a more straightforward method without the intervention of any decision-maker. In the objective weight method, each attribute is taken as an information source, and the weight of significance of each criterion reflects the amount of information contained in each of them. Finding the weight distribution of each decision criterion that has high variability and supports stable ranking is still an open research challenge. Although there are still very few studies to tackle the problem related to measuring the credibility of the weighting and ranking method [1].

However, in complex decision problems, subjective weighting can be challenging or unreliable due to its dependence on the decision-maker's knowledge, experience, and personal judgment. Objective methods provide an alternative approach to addressing these challenges. The first objective weighting method was the mean weight (MW) method, in which each decision criterion has a mean value as weight. That means each decision criterion has the same weight and demonstrates the same significance in determining alternative ranking and selection. The mean weight approach may be inadequate for complex decision problems in real-world applications, as different decision criteria hold varying levels of importance in evaluating and ranking alternatives. It is not provided any accurate model about the actual distribution of the importance of decision criteria. To further enhance the assignment of weight to each criterion in a more meaningful way, the researchers have used the concept of contrast intensity. Researchers use the concept of contrast intensity to measure the significance of decision criteria. Standard deviation (SD) is a statistical measure used to quantify the amount of variation or dispersion in a set of values, which can be applied to assess contrast intensity in decision criteria. The SD method improves upon the MW approach by using standard deviation to measure the contrast intensity of each criterion, enhancing weight differentiation. The mean in the MW method provides limited insight, as it does not accurately reflect any information regarding the distribution of data or the relative significance of the criteria. However, SD provides a higher level of variability in the data or in the distribution of criteria importance. Assigning weights to decision criteria further refines and strengthens the decision-making process through the entropy information. The entropy weighting method enhances the weight distribution of each decision criterion by applying the concept of contrast intensity, quantified through entropy-based information. Proposals for quantifying contrast intensity use the standard deviation or entropy information as the measure of importance, which is the driving force behind deriving the objective weight of each decision criterion. The entropy weighting method is highly sensitive to the evaluation of probabilities of state based on the decision matrix. The entropy weight method consists of two variants. The first employs a statistical approach, estimating state probabilities using the attribute distribution function. The second variant determines state probabilities based on the relative positioning of attributes. The entropy weighting method encounters a limitation when all the criteria being considered have identical or nearly identical values. Therefore, it is essential to determine, before applying the entropy method, how many criteria have either identical values or minimal variability among all the alternatives. When the value is the same, the entropy weighting method is not able to provide any insightful information about the significance of decision criteria to ranking alternatives. Both standard deviation and entropy techniques assign higher weights to criteria with greater variation in scores across all alternatives for a given criterion. These methods are unsuitable for assigning justifiable weights to each decision criterion and fail to provide significant additional information when all alternatives have identical or nearly identical values, i.e., when there is minimal variability. In such cases, the criteria weights derived from these methods do not provide insightful information to the decision-making process. It employs the Pearson correlation coefficient, which captures only linear relationships and lacks the capacity to address nonlinear correlations, thereby limiting its applicability to complex decision problems [2].

The author of the paper [3] has introduced a new approach, CRiteria Importance Through Intercriteria Correlation (CRITIC), that uses multi-criteria analysis through the inclusion of new dimensions. These new dimensions are termed as conflict intensity. Conflict intensity refers to the degree of disagreement among various criteria. However, if an alternative performs equally well across all decision criteria, there is no real conflict, and they may be unable to provide insightful information in the decision-making process. In such scenarios, introducing a conflict intensity dimension along with contrast intensity adds a significant amount of information and drastically alters the decision situation by providing a different way of assigning weight to each decision criterion for assessing the performance score of each alternative. Therefore, CRITIC is shown to have additional value because it considers both the contrast intensity and the conflicting relationships between criteria. The CRITIC is the enhancement of the standard deviation and entropy weighting methods, which incorporate contrast intensity along with conflict intensity. The CRITIC method provides a balanced solution that incorporates the amount of information provided by all decision criteria by considering conflict among decision criteria. However, the multi-criteria analysis result does not show the information that SD and MW weight criteria sent out, even though their values are more consistent. The author concludes that CRITIC provides a better resolution of conflict that characterizes a given decision scenario. The CRITIC method makes sure that each decision criterion is given the right amount of weight by looking at all of the criteria in a multi-criteria problem [2, 4, p. 973].

In a paper [4], a new Distance-CRITIC (D-CRITIC) weighting method has been proposed. This method has used a distance-based function to find the level of conflict when deciding the weights for the decision criteria. D-CRITIC addresses the limitations of the traditional CRITIC method, which depends exclusively on linear Pearson correlation for evaluating decision criteria. The authors of the paper [5] have proposed a new weighting method using the concept of the method based on the removal effects of criteria (MEREC). This approach assesses variations in alternative performance scores for each criterion and analyzes the impact of its removal to determine the weights. In the MEREC method, a criterion has greater weight when its removal leads to a greater effect on alternative aggregate performance. The MERCE method employs the concept of logarithmic functions to evaluate alternative performance. The performance measure function used is crucial in determining the weight of the criteria in the MEREC method. The optimization of MERE is possible by employing other functions to measure alternative functions' performance. These functions can be expressed in the form of multiplicative, exponential, or distance-based expressions [6]. The preceding discussion highlights the possibility of assigning either equal or varying weights to criteria. The equal weights involve less subjectivity, while the varying weighted distribution of decision criteria provides deeper insights about each criterion for identifying the consistent ranking and selection of appropriate alternatives. A summary of the above discussion has been mentioned in Table 5.1.

To address the challenges inherent in conventional subjective and objective weighting methods, people have proposed a hybrid version of the weighting method that is designed to determine decision criterion weights systematically and does not suffer from the problem of being too subjective or too objective by incorporating both subjective and objective features. This hybrid version of the weighting method is also known as the comprehensive weighting method. In the following section, we will look at the insights of various comprehensive weighting methods.

5.2 Need for Comprehensive Weight Method

Subjective methods such as AHP and pairwise comparison rely on expert opinion. They capture human preferences but are prone to bias and inconsistency. While on the other hand, objective methods such as entropy, CRITIC, and standard deviation use data dispersion or contrast but may ignore expert knowledge or contextual relevance. This limitation of both subjective and objective methods forces the need to integrate both subjective and objective methods, which is referred to as comprehensive methods. The comprehensive weight method integrates multiple weighting techniques to produce more reliable and unbiased results. By combining subjective (expert-based) and objective (data-driven) methods, decision-makers can achieve a balanced and justified weighting system for complex problems [7]. A comprehensive weighting method ensures that the weight is determined:

- Is balanced, considering both objective and subjective factors.
- Reduces bias and inconsistencies.
- Is robust and adaptable across different domains and decision scenarios.
- Avoids subjectivity, as relying solely on single weight assignment methods, such as expert judgment, can lead to biased outcomes.
- Improves reliability by combining multiple methods, which helps balance their respective strengths and weaknesses.

Table 5.1 Advantage and limitation of some of the subjective and objective methods

Method	Advantage	Limitations
Subjective weight method	Simple, flexible, and intuitive, leveraging human knowledge and expertise to effectively address complex ranking and optimal alternative selection problems	These methods are not suitable for most of the complex decision problems due to their dependency on the knowledge, expertise, experience, and judgment of the decision-makers
Mean weight method	It is preferred in most applications, ensuring a less subjective weight distribution for reliable decision-making	Distribute equal weight to each decision criterion that may not be suitable for complex decision problems in real-world applications. It does not provide anything accurate about the actual distribution of the importance of decision criteria
SD weight method	It provides better variability compared to mean weight and ensures a less subjective weight distribution for assessing criterion significance	Consider only contrast intensity when assigning weight to each criterion. It may not be suitable for complex decision problems. It is more sensitive to the normalization method employed
Entropy weight method	In most cases, it is better than SD, where the decision criteria of all alternatives have less variability	Similar to the SD method, only consider contrast intensity between decision criteria. It is also more sensitive to the normalization method employed. The weight obtained is highly sensitive to the entropy of various criteria. The weight obtained is highly sensitive to the distribution of the data within the domain of criteria values. It may not be very appropriate for real-world complex decision problems
CRITIC weight method	Include both contrast and conflict intensity to quantify the significance of each decision criterion	The conflict intensity uses the Pearson correlation coefficient, which handles only linear correlation among the decision criteria. It does not have the capacity to address nonlinear correlation between decision criteria. As a result, it may not be well-suited for solving complex decision problems in real-world scenarios

(continued)

Table 5.1 (continued)

Method	Advantage	Limitations
D-CRITIC weight method	Provide better weight distribution than the CRITIC method for most of the scenarios	This method used the distance correlation coefficient instead of the Pearson correlation coefficient for computing conflict intensity, similar to CRITIC. The computational complexity increases the overall time complexity of the entire method
MEREC and I-MEREC weight method	The method incorporates novel elements from other existing approaches. It provides more reliable and stable weight distribution of all decision criteria for most of the problems through assessing the performance score by the removal effect of each decision criterion	It employed the concept of the removal effect of criteria on the overall performance of alternatives. It used a logarithmic measure to determine the overall performance of the alternatives, which resulted in a nonlinear correlation

- Handles uncertainty effectively by incorporating both subjective and objective weighting approaches, offering a more holistic perspective.
- Enhances decision quality by using a well-justified weighting system, which leads to more accurate and effective outcomes.

A comprehensive weighting method is not just a technical necessity, but it is a fundamental requirement for ensuring that MADM models are accurate, transparent, adaptive, and aligned with real-world complexity and stakeholder preferences. It serves as the foundation for reliable, justifiable, and intelligent decision-making across various fields, such as engineering design, supply chain management, project management, healthcare, policymaking, and more. The comprehensive weight method offers a balanced approach by integrating expert judgment with data-driven analysis, thereby minimizing the risk of bias in the assignment of weights to each decision criterion. It ensures less reliance on a single method and provides higher robustness and reliability for critical decisions [8].

5.3 Steps in Comprehensive Weight Method

A comprehensive weight method combines multiple weighting techniques for better accuracy through adjustment proportion coefficient factor. A comprehensive weighting method may have involved the following steps:

Step 1: Select Multiple Weighting Methods
Select at least one subjective method, such as AHP or Delphi, and one objective method, such as Entropy or CRITIC.

Step 2: Calculate Weights Using Each Method
Compute weights separately for each method.

Step 3: Combine Weights Using an Aggregation Technique
Common aggregation methods include:

- Arithmetic Mean: Simple to compute but treats all methods equally, regardless of their reliability.
- Weighted Average Based on Method Reliability: Reflects confidence in each method by assigning weights (e.g., α, β), but requires predefined reliability values.
- Optimization-Based Combination: Uses mathematical programming to determine weights that minimize discrepancies between methods, offering high accuracy but involving more complex computations.

Step 4: Validate and Adjust Weights
Check for consistency by using tools like AHP's Consistency Ratio and conducting sensitivity analysis to examine how changes in weights impact the final results. The general steps involved in the computation of comprehensive weight also have been shown in Fig. 5.1.

5.3.1 Aggregation Methods

Comprehensive weighting methods integrate both subjective and objective approaches to ensure precise and balanced weighting of each decision criterion. An appropriate aggregation technique effectively leverages the strengths of both methods to achieve this. This aggregation technique used to combine subjective and objective weights is also referred to as the balance coefficient factor. The following are some commonly used balances coefficient factors:

- Linear Combination: Weights $= \alpha \times$ (Subjective Weights) $+ (1 - \alpha) \times$ (Objective Weights), where $\alpha \in [0,1]$.
- Fuzzy Comprehensive Weighting: Integrates fuzzy logic with subjective and objective methods.
- Bayesian Best–Worst Method (Bayesian BWM): Combines BWM with probabilistic modeling for uncertainty.
- AHP-Entropy Combination: Uses AHP for subjective weights and entropy for objective adjustment.

5.4 Comprehensive Weighting Methods

Comprehensive weighting methods combine both objective and subjective approaches to provide a balanced perspective that eliminates the high probability of the final decision becoming too subjective or too objective. The following may be

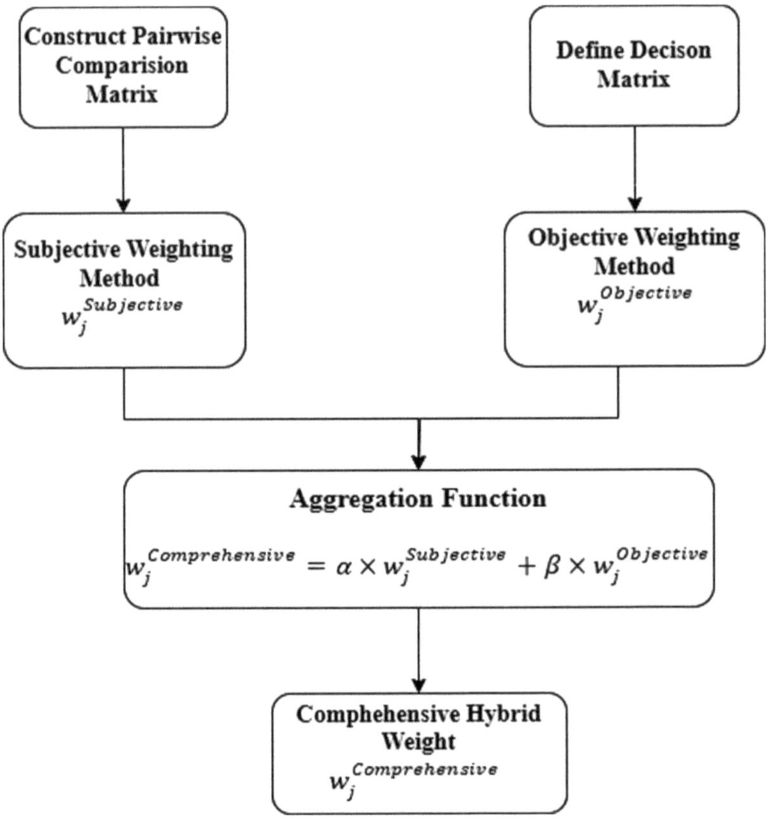

Fig. 5.1 Steps involved in comprehensive weighing method

some of the possibilities of integrating subjective and objective weighting methods to obtain a suitable and optimal comprehensive weighting method.

- Integrated AHP-Entropy Method: Combines AHP's subjective judgment with entropy's objectivity.
- Fuzzy AHP-Entropy: Uses fuzzy logic to handle uncertainty in subjective judgments and entropy for objective data analysis [9, 10].
- Best–Worst Method (BWM) + Entropy: Decision-makers define the best and worst criteria; entropy is used to fine-tune [11].
- DEMATEL–ANP Hybrid: Captures interdependencies among criteria using Decision-Making Trial and Evaluation Laboratory (DEMATEL) and Analytic Network Process (ANP) [12, 13].
- Bayesian Weighting Methods: Combine prior knowledge and data-driven insights in a probabilistic framework [14].

There are many ways to integrate subjective and objective methods through balance coefficients, as shown in Fig. 5.1. This chapter discusses only one comprehensive weighting method, which is the integration of the AHP and entropy weighting methods.

5.4.1 AHP-Entropy Weighting Method

AHP uses the concept of pairwise comparisons to assign weights to criteria based on expert judgments, which strengthens AHP in terms of capturing human expertise, priorities, and intuition, which may lead to limitations of biasness due to subjectivity. While on other hand, the entropy method uses information theory to determine the significance of each decision criterion. Entropy is used to assess how much information each decision criterion provides. The higher the variation in a criterion between alternatives, the more important it is. Entropy methods are entirely data-driven and free of human bias, but they have the limitation of potentially ignoring expert knowledge. These limitations of subjective and objective methods lead to the integration of subjective and objective methods to take advantage of both in the assignment of precise and balanced weight to each decision criterion [15]. Figure 5.2 illustrates various possible comprehensive weighting methods derived from integrating different subjective and objective approaches. These combinations aim to enhance the reliability and balance of weight assignment in multi-criteria decision-making by leveraging the strengths of both techniques.

The AHP-entropy method combines the subjective (AHP) and objective (entropy) weightings using a linear combination aggregate function, providing a balanced and comprehensive weight for each criterion. The AHP-entropy method is a comprehensive weighting approach that combines two distinct methodologies for determining the weights of each decision criteria namely AHP and entropy. This hybrid approach balances subjective expert judgment (AHP) with objective data-driven weighting (entropy), ensuring a more reliable and precise distribution of significance of each decision criteria.

AHP-entropy calculates the comprehensive weight of decision criteria through three distinct phases: first, computing the subjective weight; second, computing the objective weight; and third, integrating both subjective and objective weights using adjustment coefficient factors [16]. The AHP-entropy method has involved the following phases to obtain comprehensive weights.

- Construct Decision Matrix
- Compute Subjective Weights Using AHP
- Compute Objective Weights Using Entropy
- Combine Weights–AHP–Entropy.

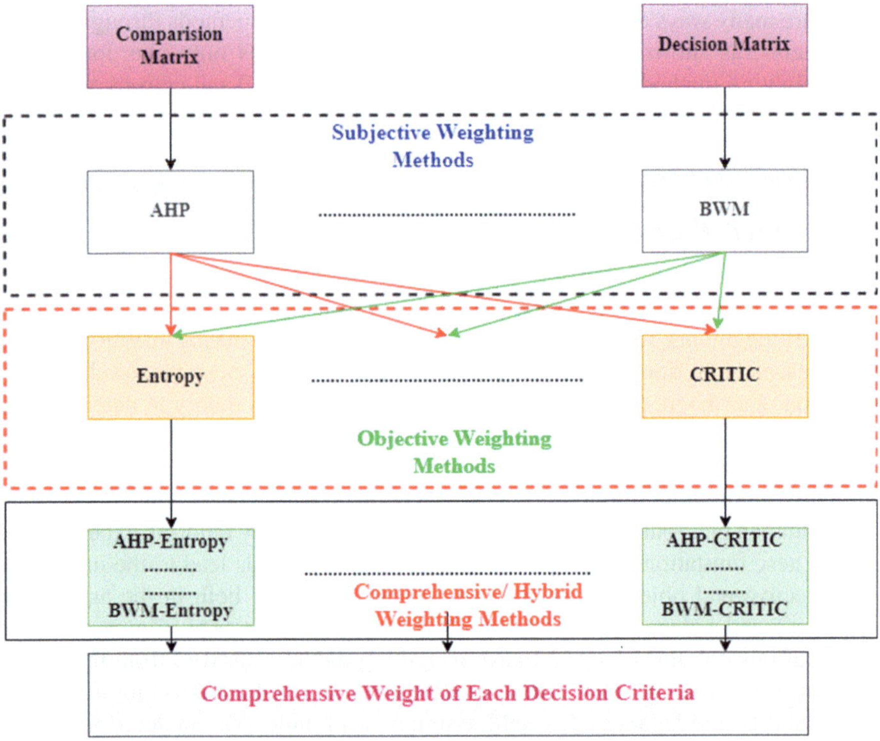

Fig. 5.2 Comprehensive weighting method obtained through the integration of different subjective and objective methods

5.4.1.1 Phase 1: Compute Subjective Weights Using AHP

The first phase of AHP–entropy is computation of subjective weight through AHP has involved the following steps:

Step 1 Define the goal first, then build a hierarchical structure consisting of criteria at the second level and alternatives at the third.

Step 2 Assess the relative importance of each criterion with respect to the objective.

Step 3 Compute the normalized comparison matrix (NCM)

$$\text{NCM} = \left[r_{i,j}\right]_{n\times n} = \left[\frac{x_{i,j}}{\sum_{i=1}^{n} x_{i,j}}\right]_{n\times n}$$

Table 5.2 Random index [17]

n	1	2	3	4	5	6	7	8	9	10
RI	0.00	0.00	0.58	0.90	1.11	1.24	1.32	1.45	1.45	1.45

Step 4 Determine the weight of each criterion by applying the given formula to the normalized comparison matrix derived from pairwise evaluations.

$$w_j^{\text{Subjective}} = \sum_{i=1}^{n} r_{i,j}$$

Step 5 Calculate weight sum value

$$\text{WSV}_i = \sum_{j=1}^{n} \frac{x_{i,j} \times w_j}{w_j}$$

$$\lambda_{\max} = \sum_{i=1}^{n} \text{WSV}_i$$

Step 6 Evaluate the consistency index (CI) to assess the logical coherence of judgments made in the pairwise comparison matrix.

$$\text{CI} = \frac{\lambda_{\max} - n}{n - 1}$$

Step 7 Calculate the consistency ratio (CR) to verify the reliability of judgments in the pairwise comparison matrix by comparing the consistency index (CI) with the appropriate random index (RI) value.

$$\text{CR} = \frac{\text{CI}}{\text{RI}}$$

The Random Index (RI) represents the average consistency index of a randomly generated pairwise comparison matrix and is presented in the Table 5.2.

5.4.1.2 Phase 2: Compute Objective Weights Using Entropy

The second phase of AHP–entropy is computation of objective weight through AHP has involved the following steps:

Step 1 Construct the normalized decision matrix by scaling the original decision values to ensure comparability across all criteria and alternatives.

$$r_{i,j} = \frac{x_{i,j}}{\sum_{i=1}^{m} x_{i,j}}$$

Step 2 Calculate the entropy value for each criterion by analyzing the normalized decision matrix, which helps measure the uncertainty and is then used to determine the degree of diversification for weighting.

$$e_j = -h\sum_{i=1}^{m} r_{i,j} \ln r_{i,j}$$

where $h = \frac{1}{\ln m}$

Step 3 Compute the weight vector

$$w_j^{\text{objective}} = \frac{1 - e_j}{\sum_{j=1}^{n} (1 - e_j)}$$

$1 - e_j$ is known as degree of diversity function.

5.4.1.3 Phase 3: Integrate Subjective and Objective Weight

The third phase of the AHP-entropy method involves combining the subjective and objective weights obtained from both approaches using an aggregation function. Various integration techniques can be applied to integrate both.

1. Simple Averaging

$$w_j^{\text{Comprehensive}} = \frac{w_j^{\text{Subjective}} + w_j^{\text{Objective}}}{2}$$

2. Weighted Combination

$$w_j^{\text{Comprehensive}} = \alpha \times w_j^{\text{Subjective}} + (1 - \alpha) \times w_j^{\text{Objective}}$$

where α is the adjustment coefficient factor such that $0 \leq \alpha \leq 1$. In the same way, other variations of the comprehensive weighting method can be developed by integrating different subjective and objective weighting methods along with taking diverse aggregation functions. These variations allow for flexible integration of expert judgment and data-driven insights. Different variations of comprehensive weighting may yield distinct sets of weight vectors for the same decision criteria, which can assign varying significance to each decision criterion.

5.5 Comparative Analysis of Comprehensive Weighting Methods

A comparative analysis of comprehensive weighting methods versus conventional subjective and objective weighting methods is essential to evaluate the robustness, accuracy, and applicability of decision-making models, especially in MCDM and MADM. The choice of weighting method depends on data availability, expert accessibility, and the need for bias mitigation. Comprehensive weighting methods are increasingly preferred in advanced MADM problems due to their balanced and robust nature. Subjective methods are useful when expert knowledge and experience are critical but may introduce bias, while objective methods are best for data-rich environments but may ignore contextual importance. The comprehensive methods provide a balanced approach that considers advantages and reduces limitations of both subjective and objective methods to enhance decision-making reliability in complex scenarios. Table 5.3 provides a comprehensive analysis that compares three categories of weighting methods in terms of methodology, advantages, disadvantages, and applicability.

5.6 Impact of Proportion Coefficient Weight Distributions

The proportion coefficient in an aggregate function plays a critical role in integrating subjective and objective weighting methods to develop a comprehensive weighting approach. This coefficient determines the relative contribution of each method in the final weight assignment for decision criteria, effectively measuring the degree of influence each method has on the outcome. Consider an aggregate function formulated as a linear combination of subjective and objective weights. Let an aggregate function be a linear combination function that integrates subjective and objective methods by involving proportion coefficients α and β.

$$w_j^{\text{Comprehensive}} = \alpha \times w_j^{\text{Subjective}} + \beta \times w_j^{\text{Objective}}$$

where $\alpha + \beta = 1$. The proportion coefficients (α and β) determine how much each weighting method influences the final decision. For instance:

- If $\alpha > \beta$, the final weights lean more on expert judgment.
- If $\alpha < \beta$, the final weights rely more on data-driven analysis.
- If If $\alpha = \beta$, both methods contribute equally.

The proportion coefficients (α and β) in the aggregate function play a decisive role in determining the influence of subjective and objective weighting methods on the final criteria weights. Their values indicate the degree to which each method contributes to the decision-making process:

Table 5.3 Comparative analysis of weighting methods in terms of basic principle, methodology, advantages, disadvantages, and applicability

Feature/aspect	Subjective methods	Objective methods	Comprehensive methods
Basis of weighting	Expert judgment	Data-driven	Hybrid (Expert + Data)
Data dependency	Low	High	Medium
Expert involvement	High	None	Medium
Bias risk	High	Low	Low to medium due to integration
Computation complexity	Low to medium depend on number of decision criteria	Medium	High
Handling qualitative criteria	Strong	Weak	Strong
Consistency check	Possible for example AHP	Not Applicable	Possible
Transparency	May lack transparency	High	Medium
Flexibility	High	Medium	High
Application area	Policy, strategy, social sciences	Engineering, operations, large data	Interdisciplinary, sustainable development
Accuracy and robustness	Subjective to judgment quality	Depends on data quality	High (balanced accuracy)
Examples	AHP, Delphi	Entropy, CRITIC	HP-entropy, fuzzy-game theory

1. If $\alpha = 0$
   The subjective weighting method is completely excluded.
   The final criteria weights depend solely on objective methods (e.g., Entropy, CRITIC).
   Useful when expert judgment is unreliable or unavailable, relying purely on data-driven analysis.
2. If $\beta = 0$
   The objective weighting method is fully disregarded.
   The final weights are determined entirely by subjective methods (e.g., AHP, Delphi).
   Applicable when data is insufficient or expert opinion is considered highly authoritative.
3. If $\alpha \neq 0$ and $\beta \neq 0$
   Both subjective and objective methods are integrated into the decision.

The final weights reflect a balanced combination of expert judgment and empirical data. Ensures robustness by mitigating biases (subjective) and contextual irrelevance (objective). When the proportion coefficient α is set to zero, it signifies that subjective weighting methods are excluded from the decision-making process, and the determination of criteria importance relies entirely on objective methods. Conversely, if coefficient β is zero, it indicates that objective weighting is omitted, and the final criteria weights are derived solely from subjective assessments. When both coefficients are non-zero, it reflects a combined approach, integrating both subjective and objective weights in the final decision. Therefore, the proportion coefficients play a crucial role in determining the extent of influence each method has in the integrated weighting process. They directly affect how decision criteria are weighted, and their values significantly influence the resulting weight vector and, consequently, the final ranking of alternatives. Therefor it can conclude that proportion coefficients function as balancing factors, regulating the trade-off between human expertise and statistical precision. Their selection must be consistent with the problem context, with objectivity prioritizing in data-rich scenarios and subjectivity when expert insight is required. A misconfigured ratio might bias results, emphasizing the importance of precise calibration (e.g., optimization techniques or sensitivity analysis) to ensure credible rankings.

5.7 Effectiveness of Various Integrated Weight Methods

The effectiveness of various hybrid weighting methods depends on the context of their application. The hybrid weighting methods combines subjective weights (expert judgment) and objective weights (entropy, CRITIC) to balance human expertise and data-driven insights. In the context of MADM, hybrid weight methods aim to balance the qualitative insights of experts (subjective weighting) with the quantitative analysis of data (objective weighting). The effectiveness of these integrated methods depends on multiple factors, including decision context, data quality, computational complexity, and stakeholder acceptance, selected subjective and objective weighting methods, aggregation function and proportion coefficient. Table 5.4 exhibits highlight of some key hybrid weighting methods and their effectiveness along with advantages, limitations, and possible general application domains [10].

5.8 Numerical Computation and Analysis

Suppose a decision-maker wants to evaluate three alternatives (A1, A2, A3) based on three criteria (C1, C2, C3). Consider a decision matrix X shown in Table 5.5 with three alternatives and three criteria. Let Y be a pairwise comparison matrix shown in Table 5.6 derived from expert judgment using the Saaty scale.

After applying the AHP-entropy method, the resulting weight vector reflects the overall importance of each decision criterion. Table 5.7 and Fig. 5.3 present a comparative analysis, highlighting the differences between subjective (AHP), objective

Table 5.4 Highlights of some key hybrid weighting methods and their effectiveness along with advantages, limitations, and application

Hybrid weighting methods	Effectiveness	Advantages	Limitations	Typical applications
AHP + Entropy	High	Simple to implementation and have balances expert opinion with data variability	AHP can suffer from consistency issues and is sensitive to the scale of judgments	Sustainability, healthcare prioritization
AHP + CRITIC [18, 19]	Very high	It considers correlation and contrast intensity while capturing the interdependence among criteria	Needs accurate data and CRITIC sensitive to outliers	Engineering design, performance evaluation
Fuzzy AHP + Entropy [20]	Very high	Handles uncertainty in expert opinion, Suitable for vague or linguistic data	Requires fuzzy logic expertise, higher complexity	Risk assessment, smart city planning
Best–Worst + Entropy [21]	High	Requires fewer comparisons, strong consistency in subjective input	Needs precise selection of best/worst criteria	Supplier selection, logistics
Delphi + Entropy [22]	Moderate to high	Reduces individual bias via group consensus, combines structured opinion with data	Time-consuming, expert panel coordination required	Policy development, technology forecasting

Table 5.5 Decision matrix

Alternative/criteria	C1	C2	C3
A1	250	3.5	50
A2	200	4.0	70
A3	300	4.2	60

Table 5.6 Pairwise comparison matrix

Citeria/criteria	C1	C2	C3
C1	1	1/3	3
C2	3	1	5
C2	1/3	1/5	1

Table 5.7 Decision criteria weight obtained from subjective (AHP), objective (Entropy) and comprehensive (AHP-Entropy) weighting methods

Criteria/methods	AHP	Entropy	AHP-entropy
C1	0.26049796	0.52409163	0.392294792
C2	0.63334572	0.11271278	0.373029249
C3	0.10615632	0.36319559	0.234675959

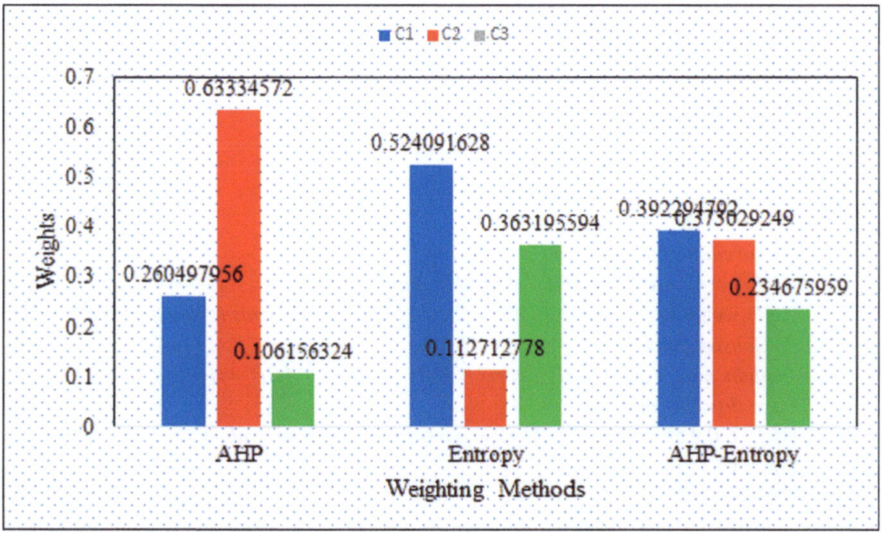

Fig. 5.3 Comparative analysis of weights of decision criteria obtained from different weighting methods

(Entropy), and comprehensive (combined) weights, thereby offering deeper insights into the relative significance of each criterion in the decision-making process.

5.9 MATLAB Implementation

This section illustrates the implementation of a hybrid weighting approach using an example of the AHP-entropy method. It demonstrates how the combined use of AHP and entropy helps determine the relative importance of decision criteria. Similarly, other hybrid weighting techniques can also be applied to evaluate criterion significance effectively. By integrating subjective and objective methods, these hybrid approaches offer a balanced assessment of criteria weights in multi-attribute decision-making problems.

```
%% Step 1: Define the decision matrix (alternatives x criteria)
% Example with 3 alternatives and 3 criteria
X = [
    250  3.5  50;   % Alternative 1
    200  4.0  70;   % Alternative 2
    300  4.2  60    % Alternative 3
];
[m, n] = size(X);  % m = alternatives, n = criteria

%% Step 2: AHP Subjective Weights (Pairwise Comparison Matrix)
% User-defined based on expert judgment
AHP_matrix = [
    1    1/3   3;
    3    1     5;
    1/3  1/5   1
];
% Normalize the AHP matrix
AHP_norm = AHP_matrix ./ sum(AHP_matrix, 1);
AHP_weights = mean(AHP_norm, 2);  % Subjective weights (column vector)
% Consistency Check
lambda_max = sum((AHP_matrix * AHP_weights) ./ AHP_weights) / n;
CI = (lambda_max - n) / (n - 1);
RI = [0 0 0.58 0.90 1.12];  % Random Consistency Index values
CR = CI / RI(n);
if CR < 0.1
    disp('AHP Consistency Ratio is acceptable.');
else
    warning('AHP Consistency Ratio exceeds threshold! Review judgments.');
end

%% Step 3: Entropy Objective Weights
% Step 1: Normalize decision matrix column-wise
P = X ./ sum(X);
% Step 2: Compute entropy for each criterion
k = 1 / log(m);
E = zeros(1, n);
for j = 1:n
    E(j) = -k * sum(P(:, j) .* log(P(:, j) + eps));  % eps avoids log(0)
end
% Step 3: Degree of diversification
d = 1 - E;
% Step 4: Entropy weights
Entropy_weights = d / sum(d);  % Row vector
```

```
%% Step 4: Combine AHP and Entropy Weights
alpha = 0.5;  % Balance coefficient: 0.5 means equal importance
Combined_weights = alpha * AHP_weights' + (1 - alpha) * Entropy_weights;

% Display results
disp('--- Subjective Weights (AHP) ---');
disp(AHP_weights');

disp('--- Objective Weights (Entropy) ---');
disp(Entropy_weights);

disp('--- Final Combined Weights (AHP–Entropy) ---');
disp(Combined_weights);
```

5.10 Summary

Comprehensive weighting plays a pivotal role in MADM, as it integrates both subjective and objective perspectives to derive more balanced and accurate decision criteria weights. The need for comprehensive weight methods arises from the limitations of relying solely on either subjective judgment, which may be biased, or objective data, which may overlook contextual nuances. Comprehensive weight methods, such as the combined use of AHP, entropy method, CRITIC, and other hybrid techniques, offer a robust solution by synthesizing different sources of weight information. A comparative analysis of these comprehensive weighting methods highlights their relative strengths and limitations in diverse decision-making contexts. The impact of proportion coefficient weight distributions is particularly significant, as they determine the extent to which each individual method influences the final integrated weight. The effectiveness of various integrated weight methods can be evaluated based on consistency, discrimination power, and sensitivity to data variations. Numerical computation and analysis provide empirical support for selecting the most appropriate method, and MATLAB implementation facilitates efficient execution and validation of these methods through automation and visualization. The comprehensive weighting methods to assign weight enhances decision quality by balancing conflicting evaluations, and its implementation. The MATLAB implementation of these methods offers a practical approach for solving complex MADM problems.

References

1. G.O. Odu, Weighting methods for multi-criteria decision making technique. J. Appl. Sci. Environ. Manag. **23**(8), 1449 (2019). https://doi.org/10.4314/jasem.v23i8.7

2. I. Mukhametzyanov, Specific character of objective methods for determining weights of criteria in MCDM problems: entropy, CRITIC and SD. Decis. Mak. Appl. Manag. Eng. **4**(2), 76–105 (2021)
3. D. Diakoulaki, G. Mavrotas, L. Papayannakis, Determining objective weights in multiple criteria problems: the critic method. Comput. Oper. Res. **22**(7), 763–770 (1995)
4. A.R. Krishnan, M.M. Kasim, R. Hamid, M.F. Ghazali, A modified CRITIC method to estimate the objective weights of decision criteria. Symmetry 2021 **13**, 973 (2021). https://www.academia.edu/download/80381553/pdf.pdf. Accessed 19 May 2025
5. M. Keshavarz-Ghorabaee, M. Amiri, E.K. Zavadskas, Z. Turskis, J. Antucheviciene, Determination of objective weights using a new method based on the removal effects of criteria (MEREC). Symmetry **13**(4), 525 (2021)
6. A.K. Yadav, K. Singh, P.K. Srivastava, P.S. Pandey, I-MEREC-T: Improved MEREC-TOPSIS scheme for optimal network selection in 5G heterogeneous network for IoT. Internet Things, 100748 (2023)
7. N. Munier, E. Hontoria, Uses and limitations of the AHP method: a non-mathematical and rational analysis, in *Management for Professionals* (Cham: Springer International Publishing, 2021). https://doi.org/10.1007/978-3-030-60392-2
8. B. Ayan, S. Abacıoğlu, M.P. Basilio, A comprehensive review of the novel weighting methods for multi-criteria decision-making. Information **14**(5), 5 (2023). https://doi.org/10.3390/info14050285.
9. D.-L. Mon, C.-H. Cheng, J.-C. Lin, Evaluating weapon system using fuzzy analytic hierarchy process based on entropy weight. Fuzzy Sets Syst. **62**(2), 127–134 (1994)
10. C. Zhong, Q. Yang, J. Liang, H. Ma, Fuzzy comprehensive evaluation with AHP and entropy methods and health risk assessment of groundwater in Yinchuan Basin, Northwest China. Environ. Res. **204**, 111956 (2022)
11. W. Xian-fa, C. Yi, S.U.N. Yue-hong, Comprehensive evaluation model of heavy metal pollution in meat products based on best-worst method and entropy method. Food Machin. **37**(4), 80–86 (2021)
12. W.-H. Tsai, W.-C. Chou, Selecting management systems for sustainable development in SMEs: a novel hybrid model based on DEMATEL, ANP, and ZOGP. Expert Syst. Appl. **36**(2), 1444–1458 (2009)
13. S. Kamranfar, Y. Azimi, M. Gheibi, A.M. Fathollahi-Fard, M. Hajiaghaei-Keshteli, Analyzing green construction development barriers by a hybrid decision-making method based on DEMATEL and the ANP. Buildings **12**(10) (2022). https://doi.org/10.3390/buildings12101641
14. F. Liu, H. Liao, Bayesian weighting for multiple criteria group decision-making with interval preference information, in *IEEE Transactions on Fuzzy Systems* (2024). https://ieeexplore.ieee.org/abstract/document/10720645/. Accessed 19 May 2025
15. R. Al-Aomar, A combined AHP-entropy method for deriving subjective and objective criteria weights
16. F. Zhou, T.-Y. Chen, A synergetic intuitionistic fuzzy model combining AHP, entropy, and ELECTRE for data fabric solution selection. Artif. Intell. Rev. **58**(5), 137 (2025). https://doi.org/10.1007/s10462-025-11128-7
17. R.W. Saaty, The analytic hierarchy process—what it is and how it is used. Math. Model. **9**(3–5), 161–176 (1987)
18. J. Qi et al., Research on the evaluation of geological environment carrying capacity based on the AHP-CRITIC empowerment method. Land **11**(8), 1196 (2022)
19. Y. Li, M. Wei, L. Liu, B. Yu, Z. Dong, Q. Xue, Evaluation of the effectiveness of VOC-contaminated soil preparation based on AHP-CRITIC-TOPSIS model. Chemosphere **271**, 129571 (2021). https://doi.org/10.1016/j.chemosphere.2021.129571
20. K. Vasantha Lakshmi, K.N. Udaya Kumara, Fuzzy MCDM techniques for portfolio selection in the post-COVID Indian mutual fund market: a comparative study of FAHP and entropy methods. J. Ambient Intell. Human Comput. **16**(1), 97–107 (2025). https://doi.org/10.1007/s12652-024-04886-9

21. Z. Tian, J. Wang, H. Zhang, An integrated approach for failure mode and effects analysis based on fuzzy best-worst, relative entropy, and VIKOR methods. Appl. Soft Comput. **72**, 636–646 (2018)
22. Q. Wang et al., Construction of the indicator framework for vaccine inclusion of routine immunization program: A Delphi-entropy method study (2023). https://www.researchsquare.com/article/rs-2772523/latest. Accessed 19 May 2025

Chapter 6
Multi-Attribute Decision-Making Models

Abstract This chapter explore the concept of MADM and fuzzy MADM ranking methods used to select the optimal alternative, derived from the weighting methods outlined in the preceding chapters Three, Four, and Five. This chapter also have comparative analyses of these ranking techniques.

Keywords Decision problem · Decision matrix · Criteria weights · Ranking · SAW · MEW · TOPSIS · VIKOR · PROMETHEE · EDAS · GRA · Fuzzy MADM

6.1 Multi-Attribute Decision Making (MADM)

Optimization issues affect almost every aspect of our lives and work. Depending on the circumstances, various methods are employed to find the best possible solution to our daily life decision problems. This requirement has prompted the development of numerous methods and algorithms over time to address a variety of challenges. One important area is MADM, whose techniques are widely used for this purpose. MADM techniques enable decision-makers to not only select the most appropriate alternative but also rank multiple alternatives in order of suitability [1]. MADM techniques are systematic approaches used when a decision must be made by evaluating a set of clearly defined alternatives based on multiple criteria or attributes. Typically, these attributes can be contradictory, meaning that improving one might negatively affect another. An MADM problem usually includes a fixed list of possible choices (alternatives), a set of decision criteria or attributes that describe what is important for the decision, the relative importance or weight assigned to each attribute, and performance ratings that show how well each alternative meets each attribute. This structured method helps decision-makers compare options thoroughly and choose the most suitable one. A solution to an MADM problem results in either identifying the single best option or producing a ranking of all available alternatives. MADM uses mathematical techniques to handle situations where decisions must be made considering multiple, often conflicting, decision criteria. It is a practical and adaptable method within the broader field of operations research, which focuses on analyzing

A. K. Yadav et al., *Multi-attribute Decision-Making: Bridging Theory and Practice*, Transactions on Computer Systems and Networks,
https://doi.org/10.1007/978-981-95-4782-1_6

and improving decision-making processes. By systematically evaluating complex factors and weighing their importance, MADM provides clear, structured information that supports decision-makers in selecting the most suitable choice. This approach ensures decisions are logical, transparent, and justifiable, even in complex scenarios [1].

6.2 Need for MADM Weighting Method

In MADM, the use of a weighting method is essential to ensure that the decision-making process accurately reflects the relative importance of each criterion involved. In most real-world scenarios, not all decision criteria carry the same level of significance; some may be more critical to the decision than others. The weighting method helps to quantify these differences by assigning appropriate values to each criterion based on their relevance, allowing decision-makers to prioritize certain attributes over others. This process not only brings clarity and structure to complex decisions but also improves the reliability and fairness of the outcome. Furthermore, weighting supports consistency, enables sensitivity analysis, and enhances the transparency of the decision-making framework by making the influence of each criterion explicit. Without proper weighting, all criteria would be treated equally, which could lead to suboptimal or misleading results [2].

The need for weight methods in MADM arises from the necessity to objectively quantify the relative importance of different criteria, ensuring a fair and structured evaluation of alternatives. Since not all criteria equally influence the final decision, weighting methods such as subjective, objective, and hybrid help assign priority values to each decision criterion, reducing subjectivity, balancing trade-offs, and improving decision accuracy. These methods enhance transparency by providing a systematic way to incorporate expert judgments or data-driven insights, making the decision-making process more reliable and reproducible across fields like engineering, business, and policy analysis. Improper weighting can undervalue critical factors, resulting in suboptimal or biased outcomes [3, 4]. The various weighting methods have been thoroughly covered in Chaps. 3, 4, and 5. In this chapter, we will focus exclusively on ranking methods, which will utilize the previously discussed weighting techniques.

6.3 MADM Methods

MADM techniques involve evaluating and ranking a finite set of alternatives based on multiple, often conflicting decision criteria. Ranking methods in MADM help decision-makers choose the best alternative(s) by comparing performance across these criteria. Let $m = \{a_{1,}a_2, a_3, \ldots a_i, \ldots a_m\}$ is set of alternatives from which one can select, $k = \{dm_{1,}dm_2, dm_3, \ldots dm_i, \ldots dm_k\}$ decision-makers can select; while

$n = \{c_1, c_2, c_3, \ldots c_j, \ldots c_n\}$ a set of attributes used to calculate each alternative's performance. The $x^k_{i,j}$ represent score of ith alternative with respect to jth criteria of decision matrix (DM) D^k, decided by kth decision-maker. The m, n, and k are total number of alternative networks, criteria, and decision-makers respectively. The $w^k = \{w^k_1, w^k_2, w^k_3, \ldots w^k_j, \ldots w^k_n\}$ is the weight vector. w^k_j is the weight of jth criteria for kth decision-maker which indicates how important to contribute to attaining the expected objective of the decision-maker to selecting the best alternative from among several alternatives [5]. The decision matrix, as defined by MADM, is used to discover and select the best alternative out of a set of alternatives.

$$D^k = \begin{bmatrix} x^k_{1,1} & \cdots & x^k_{1,n} \\ \vdots & \ddots & \vdots \\ x^k_{m,1} & \cdots & x^k_{m,n} \end{bmatrix}$$

$$w^k = \{w^k_1, w^k_2, w^k_3, \ldots w^k_j, \ldots w^k_n\}$$

Attributes of alternatives can be categorized into benefit attributes or non-benefit attributes. Beneficial attributes are those attributes with higher values desired, while non-beneficial attributes desire lower values. Non-beneficial attribute, also known as cost attribute. Let assume the set of beneficial attributes is represented as B and non-beneficial attributes as NB [6, 7]. The following section will discuss the general steps, mathematical computations, advantages, and disadvantages of some important MADM ranking methods.

6.3.1 Simple Additive Weighting (SAW) Method

It is one of the simplest MADM methods that uses a weighted average mechanism. This enables decision-makers to evaluate a score for each alternative by multiplying normalized criteria values with appropriate weights with a mathematical multiplication operation. The weighted linear combination is another name for SAW. This procedure is one of the most basic MADM methods. The total weight of each alternative in the SAW is also known as the weighted average of the network criteria, and it is this value that is used to determine the ranking of the various alternatives. It is also known as the weighted sum model. It involves the following step:

6.3.1.1 Steps

1. Define alternatives, criteria, weights.
2. Normalize the decision matrix (typically using linear normalization).
3. Multiply each normalized value by the corresponding criterion weight.

4. Sum the weighted values for each alternative.
5. Rank the alternatives based on their total scores (higher is better).

The weighted average of the ith alternative is computed by multiplying $r_{i,j}$ and its corresponding criterion weight of importance w_j, which is assigned by decision-makers. The $r_{i,j}$ is normalized score of ith alternative with respect to jth criterion in normalized DM. The best alternative is determined by summing the weighted products of all criteria for each alternative, and the alternative with the highest product weight is designated as the best alternative. The following formula can be required for the computation of the SAW ranking procedure:

$$\mathrm{PS}_i^{\mathrm{SAW}} = \sum_{j=1}^{n} r_{i,j} \times w_j$$

SAW appeals to researchers because of its simplicity, efficiency, and ability to convert raw decision data in a relatively linear process, which ensures the order of relative magnitude of standardized scores remains consistent. It is simple, intuitive, and easy to implement but more sensitive to normalization techniques and assumes linearity and independence among criteria [8–10].

6.3.2 *Multiplicative Exponent Weighting (MEW)*

It is a MADM technique used to identify the best alternative by employing the weighted product of criteria values to determine the most suitable alternative. The weighted product method is another name for the MEW. The SAW and the MEW are very similar. The main difference between MEW and SAW is that MEW uses multiplication and exponentiation operations instead of the mathematical operations of addition and multiplication. The MEW can be changed to handle the uncertainty of context better by including interval data in a smart process for choosing the right alternatives. MEW uses multiplicative aggregation of weighted criteria. It has involved the following steps.

6.3.2.1 Steps

1. Define alternatives, criteria, weights.
2. Normalize the matrix.
3. Multiply each criterion value raised to the power of its weight.
4. Compute the product for each alternative.
5. Rank alternatives (higher product is better).

The following formula can be required for the computation of the SAW ranking procedure:

$$\mathrm{PS}_i^{\mathrm{WPM}} = \prod_{j=1}^{n} r_{i,j}^{w_j}$$

There are many modifying MEW techniques that have been suggested. Some of the modifications of MEW are less prone to classification anomalies than TOPSIS. When a lower-level alternative is either removed or included in a set of alternatives, the ranking algorithm changes the ranking of the best alternatives; this phenomenon is referred to as a ranking abnormality. MEW can experience difficulties with regard to ranking anomalies. It penalizes alternatives with lower criteria scores in its ranking selection. A ranking anomaly occurred as a result of the mathematical exponential operation used in the algorithm. MEW, in contrast to SAW, exhibits nonlinear transformation characteristics. It offers advantages such as eliminating the need for additive assumptions and being more effective when criteria have different units of measurement. However, it also has limitations, including sensitivity to extremely low or high values and a lack of intuitive understanding.

6.3.3 Weighted Aggregated Sum Product Assessment Model (WASPAS)

It combines the SAW and the MEW to improve decision accuracy by balancing additive and multiplicative approaches. It is a hybrid MADM technique that effectively integrates the principles of the SAW and MEW through a balancing coefficient. By combining both additive and multiplicative aggregation approaches, WASPAS enhances the robustness and reliability of decision outcomes. In this method, alternatives are evaluated based on a set of decision criteria, each assigned a specific weight reflecting its relative importance. The final score for each alternative is computed as a weighted combination of the SAW and MEW scores, often using a balancing coefficient. It has involved the following steps.

6.3.3.1 Steps

1. Define alternatives, criteria, weights.
2. Normalize decision matrix based on beneficial and non-beneficial nature.
3. Compute performance score through SAW.
4. Compute performance score through MEW.
5. Combine performance score through a balancing coefficient.
6. Rank alternatives by final performance score.

The following formula can be required for the computation of the SAW ranking procedure:

$$\text{WASPAS} = \text{WSM} + \text{WPM}$$

$$Q_i = \lambda Q_i^1 + (1 - \lambda) Q_i^2$$

where λ is the balancing coefficient such that $0 \leq \lambda \leq 1$. Q_i^1 and Q_i^2 represent the performance score obtained from SAW and MEW methods respectively.

One of the key strengths of WASPAS is its flexibility and improved accuracy in ranking alternatives, making it particularly suitable for complex decision-making problems in fields such as engineering, supply chain management, and project selection. However, careful consideration is required in weight assignment and normalization to ensure consistent and meaningful results [11].

6.3.4 Technique for Order Preference by Similarity to Ideal Solution (TOPSIS)

Yoon and Hwang were the ones who initially proposed TOPSIS. It is a widely used MADM method. It identifies the best alternative by comparing the geometric distance of each option from an ideal (best-case) and a negative-ideal (worst-case) solution. The core idea is that the optimal alternative should be closest to the ideal solution and farthest from the negative-ideal one. This technique determines the optimal alternative based on its smallest distance from the ideal solution, also referred to as the Euclidean distance. TOPSIS is a popular MADM technique that involves evaluating the alternative that is nearest to the positive ideal solution (PIS) and farthest from the negative ideal solution (NIS). It has involved the following steps.

6.3.4.1 Steps

1. Normalize the decision matrix.
2. Multiply by weights to get the weighted normalized matrix.
3. Determine the positive ideal solution (PIS) and negative ideal solution (NIS).
4. Calculate the Euclidean distance of each alternative from PIS and NIS.
5. Compute relative closeness to the ideal solution.
6. Rank alternatives based on this closeness (higher value is better).

The information gathered from the multi-criteria decision matrix must be Euclidean normalized before being used. The decision matrix, $M = [x_{ij}]_{m \times n}$ can be used to define the multi-criteria decision problem where $a_1, a_2, \ldots a_m$ are alternatives, and $c_1, c_2, \ldots c_n$ attribute of each alternative. The entry in decision matrix x_{ij} represents score of ith alternative, with respect to jth attribute of DM. In general, vector normalization method is used to accomplish this. The following are the steps involved in obtaining the TOPSIS ranking solution to the selection problem:

Step 1 Define DM by considering the alternative and its related decision criteria.

$$\mathrm{DM} = \left[x_{ij}\right]_{m\times n}$$

Step 2 Calculation of normalization decision matrix (NDM) using vector normalization technique.

$$\mathrm{NDM} = \left[r_{i,j}\right]_{m\times n} = \left[\frac{x_{i,j}}{\sqrt{\sum_{i=1}^{m} x_{i,j}^2}}\right]_{m\times n}$$

Step 3 Compute weighted normalized decision matrix (WNDM).

$$\mathrm{WNDM} = \left[v_{i,j}\right]_{m\times n} = \left[r_{ij} \times w_j\right]_{m\times n}$$

Step 4 Calculate ideal best (v_j^+) and ideal worst (v_j^-).

$$v_j^+ = \begin{cases} v_j^{\max}, & \forall c_j \in \mathrm{B} \\ v_j^{\min}, & \forall c_j \in \mathrm{NB} \end{cases}$$

$$v_j^- = \begin{cases} v_j^{\min}, & \forall c_j \in \mathrm{B} \\ v_j^{\max}, & \forall c_j \in \mathrm{NB} \end{cases}$$

Step 5 Calculate Euclidian distance from ideal best (S_j^+) and ideal worst (S_j^-).

$$S_j^+ = \sqrt{\sum_{j=1}^{n} \left(v_{i,j} - v_j^+\right)^2}$$

$$S_j^- = \sqrt{\sum_{j=1}^{n} \left(v_{i,j} - v_j^-\right)^2}$$

Step 6 Closeness coefficient (CC_i) compute using flowing formula for each alternative.

$$CC_i = \frac{S_j^-}{S_j^- + S_j^+}$$

The solution with high C_i value is one closest to ideal solution. As a result, this is regarded as the most preferable alternative.

It has advantages, such as it considers both the best and worst scenarios, so it becomes more applicable to real-world problems. It can provide clear performance rankings and can work well with quantitative data. It is more intuitive and easier to compute. On the other hand, it has limitations, such as it is sensitive to weighting and normalization and may produce rank reversal. It requires well-defined weights. It does not have the capability to handle qualitative or fuzzy data unless extended.

The TOPSIS approach for ranking and selecting alternatives is frequently used in the literature, although it suffers from ranking irregularity. Ranking irregularity is known as ranking abnormality. A flexible and robust MADM method ensures the order of alternative rankings stays consistent when lower-ranking alternatives are discarded or added into the pool of available alternatives. Consequently, inconsistent rankings might lead to insufficient and unstable network selection decisions if a ranking algorithm has a ranking abnormality issue. In comparison to SAW and MEW, TOPSIS has a ranking abnormality problem, but it delivers higher precision in alternative ranks. Ranking abnormality is one of the important drawbacks of TOPSIS. TOPSIS is widely regarded as being intuitive to understand and computationally straightforward. Its main advantage is its flexibility, which enables decision-makers to deal with a wide range of multi-criteria problems even when the criteria are in conflict. Despite its popularity, TOPSIS is not without flaws. It is prone to rank reversal and may produce inconsistent rankings in certain circumstances. Furthermore, it does not include a method for determining or validating the weights assigned to each criterion, leaving weight selection to the decision-maker's discretion, which could compromise decision reliability [12–15].

6.3.5 Evaluation Based on Distance from Average Solution (EDAS)

The EDAS is a comparatively new and efficient MADM technique introduced by Keshavarz Ghorabaee. Initially, Keshavarz Ghorabaee used it for inventory classification. This method ranks alternatives by calculating their positive and negative distances from the average solution; specifically, an alternative with higher positive distances and lower negative distances from the average solution is considered the preferred solution. As we have already seen, TOPSIS utilizes the positive and negative ideal solutions to evaluate each alternative. This method finds a positive and negative distance between each alternative and the average alternative. The EDAS method is effective at solving inventory classification problems. EDAS is a basic and effective method of MADM, but it can only depend on decision information described by a numerical value. This means it deals with qualitative data. EDAS is very useful when we have some conflicting criteria. In this method, the average solution distance determines the best alternative. We have to weigh the desirability of the alternative against the measurement. This method is proposed to handle the problem in which the performance value of the alternative from each criterion follows the normal distribution. We have that assumption because of opting for the optimistic as well as the pessimistic score for the evaluation of alternatives and can consider the uncertainty or handle the uncertainty of decision-making data. This method has been refined for use in a variety of situations, including but not limited to, MADM problems that are subject to uncertainty.

The EDAS method is based on the idea that a good alternative should have less distance from the average solution in terms of beneficial attributes and a greater distance in terms of non-beneficial attributes. It evaluates alternatives by comparing their performance values to the average of each criterion across all alternatives. This is in contrast to methods like TOPSIS, which use the ideal and anti-ideal solutions as references, or VIKOR, which balances group utility and individual regret. It has involved the following steps.

6.3.5.1 Steps

1. Define alternatives, criteria, weights.
2. Construct the decision matrix.
3. Calculate the average solution (AV) for each criterion.
4. Calculate positive and negative distance from average (PDA & NDA) for benefit criteria (larger is better).
5. Compute weighted sums of PDA and NDA for each alternative.
6. Normalize SP and SN.
7. Calculate the appraisal score (AS).
8. Rank the alternatives based on AS; Higher AS indicates a better alternative.

The EDAS approach have the following steps to ranking and select optimal alternative.

Step 1 Define average solution matrix

$$Av_j = \left[av_{ij}\right]_{m\times n} = \left[\frac{\sum_{i=1}^{n} x_{ij}}{n}\right]_{m\times n}$$

Step 2 Determine the positive distance from average (PDA) and negative distance from average (NDA).

$$\text{PDA} = \left[pda_{ij}\right]_{m\times n} = \begin{cases} \frac{\max(0,(x_{ij}-av_{ij}))}{av_{i,j}}, & \forall c_j \in \text{B} \\ \frac{\max(0,(av_{ij}-x_{ij}))}{av_{i,j}}, & \forall c_j \in \text{NB} \end{cases}$$

$$\text{NDA} = \left[nda_{ij}\right]_{m\times n} = \begin{cases} \frac{\max(0,(av_{ij}-x_{ij}))}{av_{i,j}}, & \forall c_j \in \text{B} \\ \frac{\max(0,(x_{ij}-av_{ij}))}{av_{ij}}, & \forall c_j \in \text{NB} \end{cases}$$

Step 3 Compute weighted sum of PDA and NDA.

$$\mathrm{SP}_i = \sum_{j=1}^{n} w_j \times pda_{ij}$$

$$\mathrm{SN}_i = \sum_{j=1}^{n} w_j \times nda_{ij}$$

Step 4 Calculate the SP and SN normalized values.

$$\mathrm{NSP}_i = \frac{\mathrm{SP}_i}{\max_i(\mathrm{SP}_i)}$$

$$\mathrm{NSN}_i = 1 - \frac{\mathrm{SN}_i}{\max_i(\mathrm{SN}_i)}$$

Step 5 Normalize the value of NSP_i and NSN_i.

$$\mathrm{AS}_i = \frac{1}{2}(\mathrm{NSP}_i + \mathrm{NSN}_i)$$

The EDAS method offers several significant advantages that make it appealing for MADM. One of its primary strengths is that it does not require normalization of the decision matrix. This means that the method can operate directly on the raw data values without the need to transform them into a standardized scale, thereby reducing computational complexity and minimizing the risk of distortion introduced by normalization techniques. Additionally, EDAS is less sensitive to extreme or outlier values compared to methods like TOPSIS or VIKOR, which rely heavily on identifying the best and worst values (ideal and anti-ideal solutions). By using the average value of each criterion as a reference point instead of the highest and lowest values, EDAS provides a more stable and fair evaluation, especially in datasets where outliers can affect the results. The interpretation of results in EDAS is also relatively straightforward. It evaluates how far each alternative is from the average solution, positively or negatively, making the analysis more intuitive and transparent to decision-makers. Furthermore, EDAS is versatile in handling both types of criteria: beneficial (where higher values are preferred) and non-beneficial (where lower values are desired), which increases its applicability across various decision-making scenarios.

However, EDAS is not without limitations. Its reliance on the average value as a reference point can sometimes be a drawback. The average does not necessarily reflect the best performance or the most desirable target; as such, EDAS may not adequately highlight alternatives that significantly outperform or underperform in comparison to the best or worst options. This makes it less effective in capturing the extremity of performance, which is a strength of ideal-based methods like TOPSIS and VIKOR. Moreover, EDAS employs a linear weighting scheme, where the importance of each criterion is aggregated linearly. This approach assumes that criteria are independent and does not consider possible interdependencies or interactions between them, which can be a significant limitation in complex, real-world decision-making problems where such relationships exist [16–18].

6.3.6 VlseKriterijumska Optimizacija I Kompromisno Resenje (VIKOR)

The VIKOR method, developed by Serafim Opricovic in 1998, is a MADM method that focuses on ranking and selecting from a set of alternatives in the presence of conflicting criteria. It introduces the concept of a compromise solution, which is the closest to the ideal solution (best among all criteria) and is suitable when the decision-maker wants to reach a consensus solution or when there is a need to rank alternatives and select the best one under conflict resolution principles. The VIKOR method is a popular MADM technique used to rank and select the best alternative among a set of options with conflicting criteria. It focuses on finding a compromise solution that is closest to the ideal while considering both the best and worst possible outcomes. It emphasizes the maximum group utility (majority rule) and minimum individual regret of the opponent (opposing viewpoint). This approach is perfect for non-compensatory issues that do not require any trade-offs between criteria. It uses linear normalization of the decision matrix and provides rankings and compromise solutions. It has involved the following steps.

6.3.6.1 Steps

1. Define alternatives, criteria, weights.
2. Determine best and worst values.
3. Compute the normalized measure through utility measure and regret measure. These measure the overall deviation and the maximum deviation, respectively, from the ideal.
4. Compute the VIKOR Index.
5. Rank alternatives according to utility measure, regret measure, and VIKOR index.
6. Propose compromise solution.

It is an important multicriteria and compromise solution. A compromise solution means an agreement established by mutual consideration. Serafim Opricović was first proposed VIKOR as a compromise method to calculate the preference ranking score of a single alternative with numerous different criteria. The VIKOR ranking score is based on the opponent's highest group utility as well as the least amount of individual regret. The decision matrix, $D(m \times n)$ can be used to define the multi-criteria decision problem where $a_1, a_2, \ldots a_m$ are alternatives, and $c_1, c_2, \ldots c_n$ attribute of each alternative. The entry in decision matrix x_{ij} represents score of ith alternative, with respect to jth attribute of DM. The VIKOR may be implemented using the steps below:

Step 1 For each attribute, find the best and worst values for each alternative.

$$x_j^+ = \begin{cases} x_j^{\max}, \ \forall c_j \in \mathrm{B} \\ x_j^{\max}, \ \forall c_j \in \mathrm{NB} \end{cases}$$

$$x_j^- = \begin{cases} x_j^{\min}, \ \forall c_j \in \mathrm{B} \\ x_j^{\max}, \ \forall c_j \in \mathrm{NB} \end{cases}$$

Step 2 Compute unity measure S_i.

$$S_i = \sum_{i=1}^{m} \left(w_i \times \frac{x_j^+ - x_{i,j}}{x_j^+ - x_j^-} \right)$$

Step 3 Calculate individual regrate R_i.

$$R_i = \max_j \left(w_j \times \frac{x_j^+ - x_{i,j}}{x_j^+ - x_j^-} \right)$$

Step 4 Compute

$$S^+ = \min_i S_i$$

$$S^- = \max_i S_i$$

$$R^+ = \min_i R_i$$

$$R^- = \max_i R_i$$

Step 5 Compute

$$Q_i = \vartheta \times \frac{S_i - S^+}{S^- - S^+} + (1 - \vartheta) \times \frac{R_i - R^+}{R^- - R^+}$$

The variable ϑ is introduced as a weight in order to recognise and reward the strategy that maximises group utility. Alternative with minimum value ranked as one.

Accepting advantage and acceptable stability in decision-making are considered in this compromise solution.

Case 1: Acceptable advantage

$$Q(A^2) - Q(A^1) \geq DQ$$

$$DQ = \frac{1}{m-1}$$

where A^1 and A^2 is the alternative score of rank one and rank two.

Case 2: Acceptable stability in decision-making if one of the basic requirements is not encountered, a list of compromise solutions is provided, which includes the following options:

1. Alternative a^1 and a^2 if only condition C_2 is not satisfied, or
2. Alternative a^1, a^2, ... $a^{(m)}$ if condition C_1 is not satisfied and $a^{(m)}$ is computed by the relation

$$Q(a^m) - Q(a^1) < DQ$$

VIKOR is a powerful MADM method for finding a compromise solution that balances the interests of different stakeholders and considers conflicting criteria in decision-making processes. It offers advantages such as balances between group utility and individual regret and is more suitable for problems with conflicting criteria. It does not require criteria to be of the same nature (benefit or cost) and provides a structured compromise solution. On the other hand, it also has some limitations, such as it is more sensitive to the value of the weight of the strategy of the majority of criteria. Typically, decision-makers set the weight of the strategy at 0.5. It requires decision criteria to be precisely weighted and assumes linearity in normalization. It is not ideal when decision-makers prefer pure utility or pure regret approaches [19].

6.3.7 Preference Ranking Organization Method for Enrichment Evaluation (PROMETHEE)

PROMETHEE is a well-known MADM method designed to assist in complex decision-making where multiple, often conflicting, criteria are involved. Developed by Jean-Pierre Brans and Bertrand Mareschal in the early 1980s, PROMETHEE is especially valued for its transparency, simplicity in implementation, and ability to incorporate decision-makers' preferences. PROMETHEE is part of the outranking methods in MADM, where alternatives are compared pairwise across different criteria rather than being aggregated into a single utility score. It allows decision-makers to express preferences over criteria and provides partial or complete rankings of alternatives [20–22]. PROMETHEE is particularly useful when:

- Decision-makers need to compare alternatives based on quantitative and qualitative criteria.
- There is a need to incorporate preference functions to model how much one alternative is preferred over another.
- Partial and complete rankings are required for better decision transparency.

In PROMETHEE for each decision criterion, a preference function is chosen to determine how much one alternative is preferred over another [23]. Six standard preference functions are available [22]:

Usual Criterion (No threshold, strict preference)
Quasi-Criterion (Indifference threshold)
Linear Preference (Preference threshold)
Level Criterion (Indifference and preference)
Linear Preference with Indifference
Gaussian Criterion (Smooth preference).

These allow flexibility in modeling preferences, including thresholds for indifference and strict preference. It is one of the well-known flexible and easy to implement MADM approaches that uses the outranking concept. It has various versions. The preference function is an essential part of PROMETHEE [24]. Using the preference function [25], we may convert the difference between two alternatives into a preference degree with values ranging between [0, 1] for each criterion that is specified. The decision matrix, $D(m \times n)$ can be used to define the multi-criteria decision problem where $a_1, a_2, \ldots a_m$ are alternatives, and $c_1, c_2, \ldots c_n$ attribute of each alternative. The entry in decision matrix $x_{i,j}$ represents score of ith alternative, with respect to jth attribute of DM. In this section, we have discussed only PROMETHEE-I and PROMETHEE-II. The PROMETHEE algorithm is implemented in the following steps, which are detailed below.

Step 1 Normalize evaluation matrix (Decision matrix)

$$R_{i,j} = \begin{cases} \frac{[x_{i,j} - \min(x_{i,j})]}{[\max(x_{i,j}) - \min(x_{i,j})]}, & \forall c_j \in \mathrm{B} \\ \frac{[\max(x_{i,j}) - x_{i,j}]}{[\max(x_{i,j}) - \min(x_{i,j})]}, & \forall c_j \in \mathrm{NB} \end{cases}$$

Step 2 Compute evaluative difference of ith alternative with respect to another alternative.

Step 3 Preference function computation can be done by

$$P_j(a,b) = \begin{cases} 0, & \text{if}\, R_{aj} < R_{bj} \Rightarrow D(M_a - M_b \le 0) \\ R_{aj} - R_{b,j}, & \text{if}\, R_{a,j} > R_{b,j} \Rightarrow D(M_a - M_b > 0) \end{cases}$$

Step 4 Calculate the aggregated preference aggregate preference function,

$$\prod(a,b) = \frac{\left[\sum_{j=1}^{n} w_j \times P_j(a,b)\right]}{\sum_{j=1}^{n} w_j}$$

Step 5 Find the leaving (positive) flow for ath alternative φ^+ and entering (negative) flow φ^- while $a \neq b$.

$$\varphi^+ = \frac{1}{m-1}\sum_{b=1}^{m}\prod(a,b)$$

$$\varphi^- = \frac{1}{m-1}\sum_{b=1}^{m}\prod(b,a)$$

Step 6 Determine the net outranking flow for each alternative.

$$\varphi(a) = \varphi^+(a) - \varphi^-(a)$$

Step 7 Compute ranking of all assumed alternative based on value of $\varphi(a)$. High value of $\varphi(a)$ rank as heighest and low value $\varphi(a)$ ranked lowest.

In case of PROMETHEE I technique we are taking total value of leaving (φ^+) and entering (φ^-) instead of taking the average of leaving and entering flow while $a \neq b$

$$\varphi^+ = \sum_{b=1}^{m}\prod(a,b)$$

$$\varphi^- = \sum_{b=1}^{m}\prod(b,a)$$

In PROMETHEE I instead of defining rank, preference is computed and decision taken as.

1. Better the leaving flow value better is the alternative.
2. Lower is the entering flow value better the alternative.

Preference indicator
Alternative a is preferred over alternative b, ${}_aP_b$ if

$$\varphi^+(a) > \varphi^+(a) \text{ and } \varphi^-(a) - \varphi^-(a) \text{ Or}$$

$$\varphi^+(a) > \varphi^+(a) \text{ and } \varphi^-(a) = \varphi^-(a) \text{ Or}$$

$$\varphi^{+}(a) = \varphi^{+}(a) \text{ and } \varphi^{-}(a) < \varphi^{-}(a)$$

Indifference situation, ${}_aI_b$ if

$$\varphi^{+}(a) = \varphi^{+}(a) \text{ and } \varphi^{-}(a) = \varphi^{-}(a)$$

Incomparable situation, ${}_aR_b$ if

$$\varphi^{+}(a) > \varphi^{+}(a) \text{ and } \varphi^{-}(a) > \varphi^{-}(a) \text{ Or}$$

$$\varphi^{+}(a) < \varphi^{+}(a) \text{ and } \varphi^{-}(a) < \varphi^{-}(a)$$

Several variants enhance PROMETHEE's applicability:

PROMETHEE I (Partial Ranking): Allows incomparability.

PROMETHEE II (Complete Ranking): Full ranking based on net flow.

PROMETHEE III (Interval Ranking): Introduces uncertainty via intervals.

PROMETHEE IV (Continuous Case): Extends to infinite alternatives.

PROMETHEE V (Optimization with Constraints): Integrates with MCDA under constraints.

PROMETHEE GDSS (Group Decision Support): Aggregates multiple decision-makers' preferences.

PROMETHEE is a powerful and flexible MADM method that provides structured decision-making under multiple criteria. Its ability to incorporate different preference functions makes it adaptable to various real-world problems. However, careful consideration must be given to weight assignment and preference function selection to ensure robust results. It offers the advantage of being intuitive and transparent, which means it is easy to explain to decision-makers. It can customize preference functions to capture real-world preferences more realistically. It is more suitable for quantitative and qualitative data. It also provides interactive visual tools that often used with the GAIA plane, which provides graphical representations. On the other hand, it also has some crucial limitations and considerations. It requires precise weight assignments, which can introduce subjectivity. It is more sensitive to preference function selection. It does not inherently handle group decision-making unless adapted, and scalability can be an issue for very large datasets [26–29].

6.3.8 Grey Relational Analysis (GRA)

GRA is a MADM technique that helps in evaluating and selecting the best alternative from a set of options, considering multiple criteria. It focuses on the grey system, meaning systems with incomplete, uncertain, or fuzzy information, often encountered in real-world decision-making scenarios. GRA identifies the degree of relationship between different elements in a system, allowing for a more comprehensive assessment of alternatives, particularly when dealing with imprecise data. GRA is a widely used technique within the MADM domain, particularly useful when dealing with systems characterized by incomplete, uncertain, or imprecise information. It belongs to the broader family of Grey system theory introduced by Julong Deng in the early 1980s. Grey system theory deals with systems that have partially known and unknown information. It lies between a white system that is fully known and a black system that is completely unknown. GRA evaluates the degree of similarity or relational grade between sequences (alternative performance against criteria) by analyzing the geometric proximity to an ideal reference [30, 31].

GRA is a technique used to aggregate individual rankings from multiple decision-makers into a consensus ranking while minimizing disagreement. Below is a detailed breakdown of GRA-based MADM methods. Different decision-makers may have conflicting rankings. GRA seeks to minimize rank disagreement while preserving fairness. It is ranking-oriented and aims to find a compromise ranking. Imagine a scenario where a company needs to select a new supplier based on several criteria like price, quality, and delivery time. Using GRA, the company could evaluate each supplier's performance against these criteria and determine the grey relational degree between each supplier and the ideal supplier (a hypothetical supplier with the best performance on all attributes). The supplier with the highest grey relational degree would be considered the best option.

It was proposed by Deng Julong in 1989 and also term as grey system theory. It reflects the system science and uncertainty system theory concepts. GRA is comprised of three components: grey relational coefficient, data processing, and grey relational grade. A common use of GRA is to turn a multi-objective optimization issue into a single-objective optimization problem [32]. Gray system theory is used in a MADM approach. It is used to look at the relationship between a reference and its comparison series. All of the performance scores of each alternative are used to create the comparison series. The comparison series aids in the elimination of scale mistakes in decision-making that can arise when comparing criteria of varying dimensions units or scales, as well as varying ranges. The comparison series of each alternative is compared to a given reference series to obtain the gray coefficient of the relative estimate for the compared alternatives. GRA provides a valuable tool for MADM by offering a robust and flexible approach to evaluating alternatives in situations where information is uncertain or incomplete, making it suitable for a wide range of decision-making problems. In order to determine how closely comparative sequence is related to reference sequence, GRA technique is applied. Therefore, the most advantageous alternative is the one that is most similar to the reference sequence. Although uncertainty may also be dealt with using fuzzy theories. In some bad data circumstances, GRA's capacity to handle missing information is more realistic and effective. Another key benefit of GRA is its capacity to produce satisfactory results with a small amount of data or a high degree of unpredictability in the decision criteria [33–35]. It is also used for optimal multi-objective problems. It has involved the following steps.

6.3.8.1 Steps

1. Define alternatives, criteria, weights.
2. Normalize the decision matrix to transforms the data into a dimensionless matrix between 0 and 1.
3. Determine the reference sequence that is the ideal reference sequence which is usually the best value in each criterion.
4. Calculate Grey relational coefficient between reference and each alternative.
5. Calculate Grey relational grade.
6. Ranking the alternatives based on descending order of grey relational grade; higher value means closer to the ideal.

The decision matrix, $D(m \times n)$ can be used to define the multi-criteria decision problem where $a_1, a_2, \ldots a_m$ are the alternatives, and $c_1, c_2, \ldots c_n$ attribute of each alternative. The entry in decision matrix $x_{i,j}$ represents score of ith alternative, with respect to j^{th} attribute of DM. GRA may be used to choose the best alternative from a variety of possibilities based on many factors by following the procedures below:

Step 1 Normalize the given decision matrix using linear min max normalization method

$$x_i^* = \begin{cases} \frac{x_i^0(k) - \min(x_i^0)}{\max(x_i^0) - \min(x_i^0)}, & \forall c_j \in \mathrm{B} \\ \frac{\min(x_i^0) - x_i^0(k)}{\max(x_i^0) - \min(x_i^0)}, & \forall c_j \in \mathrm{NB} \end{cases}$$

Step 2 Grey relational coefficient

$$\zeta_i(k) = \frac{\Delta_{\min} + \xi \Delta_{\max}}{\Delta_{0i}(k)^\circ\mathrm{C} + \xi \Delta_{\max}}$$

Deviation sequence

$$\Delta_{0i} = x_0^*(k) - x_i^*(k)$$

$$\Delta_{\min} = \min_{\forall j \in i} \max_{\forall k} x_0^*(k) - x_i^*(k)$$

$$\Delta_{\max} = \max_{\forall j \in i} \min_{\forall k} x_0^*(k) - x_i^*(k)$$

where Δ_{0i} is the deviation sequence. A value of ξ is smaller and the distingued ability is the longer, $\xi = 0.5$ is generally used. ξ is distinguished coefficient.

Step 3 Grey relation grade

$$\gamma_i = \frac{1}{n} \sum_{k=1}^{n} w_k(k) \xi_k(k)$$

GRA techniques in MADM provide a structured way to combine multiple rankings into a fair consensus. While computationally challenging, they are highly valuable in group decision-making scenarios where ordinal preferences matter. GRA has the advantage that it can handle uncertain, incomplete, and imprecise data well. It is a simple mathematical structure with strong interpretability and does not require large data samples. It can be applied to qualitative and quantitative criteria and more suitable for dynamic systems with time-varying data. On the other hand, it has few limitations. It is more sensitive to the distinguishing coefficient. It requires careful normalization to avoid distortion and assumes linearity in relational grades. It asks about decision criteria weights that must be assigned, which may introduce subjectivity unless derived objectively. It is computationally intensive for large alternatives and does not provide cardinal information (loses strength of preference). It is more susceptible to strategic manipulation [36, 37].

6.4 Fuzzy MADM Methods

The traditional MADM represents a critical approach in operations research and management science, dealing with the evaluation and selection of alternatives based on multiple, often conflicting criteria. The conventional MADM methods operate under several restrictive assumptions that limit their real-world applicability:

1. Precision Requirement: They demand exact numerical values for all criteria weights and alternative ratings.
2. Binary Logic: They typically use crisp true/false dichotomies that do not reflect human reasoning.
3. Static Framework: They often fail to account for the dynamic nature of real-world decision parameters.
4. Subjectivity Handling: They provide poor mechanisms for incorporating expert judgment and qualitative factors.

These limitations become particularly problematic in complex scenarios such as large-scale infrastructure projects, environmental impact assessments, healthcare system evaluations, military strategy formulation, and supply chain network design [38–40].

Fuzzy set theory, introduced by Lotfi Zadeh in 1965, provides a mathematical framework for dealing with vagueness and uncertainty inherent in human cognition and real-world phenomena. By extending conventional set theory to allow partial membership (values between 0 and 1), fuzzy logic enables more nuanced representation of decision parameters. Key characteristics that make fuzzy logic suitable for complex MADM:

1. Gradual Transitions: Allows representation of "more or less" rather than "either/or"
2. Linguistic Variables: Enables use of natural language terms for example, "high," "medium," "low")

3. Uncertainty Incorporation: Can model imperfect or incomplete information
4. Human-Centric: Better aligns with how experts actually think and express judgments.

The integration of fuzzy logic with traditional MADM enable decision more precise and applicable where decisions must account for uncertainty, imprecision, and subjective judgments, traditional MADM methods often prove inadequate. This integration may provide following features:

1. Handling Qualitative and Quantitative Decision Criteria Simultaneously: Complex decisions typically involve both measurable such as cost, time, and qualitative such as aesthetic value, political acceptability aspects. Fuzzy MADM provides unified framework for converting linguistic assessments into mathematical representations and maintaining the uncertainty inherent in qualitative judgments throughout the analysis. Fuzzy MADM methods ensuring all criteria receive appropriate consideration regardless of measurability.
2. Managing Information Imperfection: In deep complexity, decision-makers face issues like missing data that is unavailable or too costly to obtain, conflicting information from different sources, varying reliability of information sources and non-stationary data that is parameters that change during analysis. In such scenarios fuzzy MADM methods can represent information reliability as degrees of membership and incorporate confidence levels in expert judgments FMADM methods may use interval-valued or type-2 fuzzy sets for additional uncertainty dimensions and apply fuzzy clustering to handle conflicting data sources.
3. Accommodating Multiple Stakeholder Perspectives: High-stakes decisions typically involve diverse stakeholders with different value systems, varying risk tolerances, conflicting priorities, and distinct linguistic expressions of preferences. In such scenarios the fuzzy approaches enable aggregation of heterogeneous preference structures and representation of consensus degrees. It provides modeling of compromise solutions and visualization of solution spaces rather than single-point answers.
4. Addressing Dynamic Decision Environments: Many complex systems exhibit time-varying parameters, feedback loops, emergent properties, and path dependencies. In such scenarios fuzzy MADM extensions can incorporate temporal fuzzy sets for changing criteria weights and use fuzzy cognitive maps to model system dynamics. FMADM approach apply adaptive fuzzy inference for learning systems and implement fuzzy Markov processes for sequential decisions
5. Dealing with Large-Scale, High-Dimensional Problems: As problem dimensionality increases (many alternatives × many criteria × many scenarios), fuzzy methods help by enabling hierarchical decomposition with fuzzy aggregation and supporting fuzzy dimensionality reduction techniques. The FMADM approach allowing fuzzy rule-based screening of alternatives and facilitating fuzzy clustering of similar alternatives.

The following may be the cause of the integration of fuzzy logic with the traditional MADM approach.

1. Handling Uncertainty and Vagueness: Real-world decision-making often involves qualitative, imprecise, or subjective data for example high quality, moderate risk. Fuzzy MADM uses fuzzy sets to model such linguistic variables and uncertainty effectively.
2. Inadequacy of Crisp Values: Classical MADM methods like AHP, TOPSIS, or ELECTRE typically rely on precise numerical inputs. However, decision-makers may not always be able to provide exact values. Fuzzy MADM allows for approximate inputs using fuzzy numbers or linguistic terms.
3. Complexity of Human Judgment: Human preferences are often vague and not strictly quantifiable. Fuzzy MADM supports a more natural representation of human reasoning, enabling better alignment with how decisions are actually made.
4. Robustness in Multi-criteria Analysis: In multi-criteria problems where trade-offs are involved, fuzzy MADM enhances robustness by providing flexible weighting and evaluation under imprecise conditions.
5. Wide Range of Applications: Fields like supply chain management, healthcare, environmental planning, engineering design, and project selection benefit from fuzzy MADM due to the inherent uncertainty in data and preferences.
6. Improved Decision Quality: By integrating fuzziness, the decisions are more realistic and inclusive of uncertainty, leading to outcomes that are more acceptable and justifiable.

In an increasingly complex world where decisions must account for uncertainty, imprecision, and subjective judgments, traditional MADM methods often prove inadequate. This is where fuzzy MADM emerges as an essential tool, particularly in high-stakes, complex decision environments. The need for Fuzzy MADM (FMADM) arises due to the limitations of classical decision-making methods when dealing with uncertainty, vagueness, and imprecision in real-world scenarios. FMADM addresses these challenges by incorporating fuzzy logic, allowing for the representation, and handling of vague or imprecise information. In an increasingly complex and uncertain world, fuzzy MADM has become not just useful but essential for high-quality decision-making. By properly accounting for the inherent vagueness, subjectivity, and dynamism of real-world problems especially those of great depth and consequence fuzzy approaches provide more realistic, flexible, and human-compatible decision support than conventional methods. As challenges grow in scale and complexity across all domains, the need for sophisticated fuzzy MADM techniques will only intensify, driving continued methodological innovation and practical application. The depth of modern decision problems demands nothing less than the nuanced, robust capabilities that fuzzy MADM provides. Organizations facing complex choices would do well to build competency in these methods to maintain decision quality in the face of uncertainty and complexity. FMADM provides a more robust and flexible framework for decision-making in situations where uncertainty and imprecision are significant factors, leading to more accurate and realistic decision outcomes. In this we have discusses only one fuzzy TOPSIS FMADM approach [41–43].

6.4.1 Fuzzy TOPSIS

Fuzzy TOPSIS is an extension of the classical TOPSIS method, which incorporates fuzzy set theory to handle uncertainty and imprecision in decision-making. It is widely used in MADM problems where input data is expressed using linguistic terms or fuzzy numbers. The fuzzy TOPSIS method has the capability to adopt real-world data that often comes with vagueness, for example, in terms of high, medium, and low. It can be work as experts that may have subjective or approximate opinions. In this, the fuzzy logic component allows modeling this imprecision with fuzzy numbers, usually triangular or trapezoidal fuzzy numbers. It is a technique used in situations where the performance values in the decision matrix are expressed as linguistic terms rather than precise numerical values, as decision-makers find it challenging to assign exact rankings. Since assigning ranks without any numeric values is difficult, decision-makers use fuzzy AHP to derive fuzzy numbers for the weights of each criterion instead of directly assigning linguistic terms to them [44–46]. The concept of group decision-making has two or more decision-makers involved. These decision-makers may have the same or different importance. It has involved the following steps.

1. Define the decision problem and construct the fuzzy decision matrix.
2. Normalize the fuzzy decision matrix.
3. Compute the weighted fuzzy decision matrix through multiply normalized fuzzy ratings by fuzzy weights.
4. Determine fuzzy positive ideal solution (FPIS) and fuzzy negative ideal solution (FNIS).
5. Calculate the distance of each alternative from FPIS and FNIS.
6. Compute closeness coefficient (CC).
7. Rank the alternatives.

The mathematical step-by-step computation of the performance score of each alternative and selection of ranking involved in fuzzy TOPSIS can be described as follows.

Step 1 Convert single decision matrix from all k decision matrix

$$\overline{x_{i,j}} = \left(\min_k \left\{ a_{i,j}^k \right\}, \frac{1}{k} \sum_{k=1}^{k} b_{i,j}^k, \max_k \left\{ c_{i,j}^k \right\} \right)$$

In case if decision-maker take different fuzzy weight for criteria can calculate by combined fuzzy weight for combined decision matrix by using

$$w_{j,i} = \left(\min_k \left\{ w_{i,j}^k \right\}, \frac{1}{k} \sum_{k=1}^{k} w_{i,j}^k, \max_k \left\{ w_{i,j}^k \right\} \right)$$

Step 2 Compute normalized fuzzy DM

$$\overline{r_{i,j}} = \begin{cases} \frac{a_{i,j}}{c_j^*}, \frac{b_{i,j}}{c_j^*}, \frac{c_{i,j}}{c_j^*} \text{ and } c_j^* = \max_i\{c_{i,j}\}, \ \forall c_j \in \text{B} \\ \frac{a_j^-}{c_{i,j}}, \frac{a_j^-}{c_{i,j}}, \frac{a_j^-}{c_{i,j}} \text{ and } a_j^- = \min_i\{a_{i,j}\}, \ \forall c_j \in \text{NB} \end{cases}$$

Step 3 Computer the weighted normalized fuzzy DM

$$\overline{v_{i,j}} = \overline{r_{i,j}} \times w_j$$

Step 4 Compute fuzzy PIS and fuzzy NIS

$$A^* = \left(\overline{v}_1^{,*}, \overline{v}_2^{,*}, \overline{v}_3^{,*} \ldots \overline{v}_n^{,*}\right)$$

where $\overline{v}_n^{,*} = \max_i\{v_{ij3}\}$

$$A^- = \left(\overline{v}_1^{,-}, \overline{v}_2^{,-}, \overline{v}_3^{,-} \ldots \overline{v}_n^{,-}\right)$$

where $\overline{v}_n^{,-} = \max_i\{v_{ij1}\}$

Step 5 Compute the distance from each alternative to FPIS and FNIS

$$d(\overline{x},\overline{y}) = \sqrt{\frac{1}{3}\left[(a_1 - a_2)^2 + (b_1 - b_2)^2 + (c_1 - c_2)^2\right]}$$

Step 6 Compute the distance from each alternative to the FPIS and to FNIS

$$d_i^* = \sum_{j=1}^{n} d\left(\overline{v_{i,j}}, \overline{v}_j^{,*}\right)$$

$$d_i^- = \sum_{j=1}^{n} d\left(\overline{v_{i,j}}, \overline{v}_j^{,-}\right)$$

Step 7 Compute closeness coefficient CC_i for each and every alternative

$$CC_i = \frac{d_i^-}{d_i^- + d_i^*}$$

The fuzzy TOPIS technique has various advantages, including handling uncertainty and vagueness effectively and the capability to work with both numerical and linguistic data. This approach can provide a systematic ranking of alternatives and is flexible with different fuzzy number types such as triangular and trapezoidal. On the other hand, irrespective of this advantage, it also has some limitations and challenges. Its computational complexity increases with more criteria/alternatives. It requires expert knowledge for fuzzy rating assignments and sensitivity to weight determination methods. Fuzzy TOPSIS is a powerful extension of traditional TOPSIS, enabling decision-makers to handle imprecise, uncertain, and subjective data effectively. By integrating fuzzy logic, it provides a more realistic and flexible approach to complex decision-making problems across various industries. For researchers and practitioners, mastering Fuzzy TOPSIS can significantly enhance decision accuracy in environments where crisp data is unavailable or unreliable [47, 48].

6.5 Impact of Comprehensive Weight Methods Ranking Order

Ranking systems are widely used in decision-making, evaluation, and prioritization across various fields such as economics, engineering, management, and social sciences. A critical aspect of ranking is the assignment of weights to different criteria, as these weights significantly influence the final ranking order. In MADM, assigning weights to each decision criteria is one of the most pivotal steps. These weights influence the final decision by indicating the relative importance of each criterion. However, a significant challenge arises when different weight determination methods yield different priority orders, impacting the final ranking of alternatives. This issue gave rise to comprehensive weight methods, which aim to integrate multiple weighting strategies both objective and subjective to obtain balanced and robust rankings. Understanding their impact on ranking order is crucial for improving the reliability and consistency of MADM results [49].

Subjective methods reflect human judgment, making them useful when expert knowledge is critical. However, they may introduce bias due to individual knowledge, expertise, and preferences, leading to inconsistent rankings if experts disagree. On other hand objective methods eliminate human bias, making them more reliable for data-driven decisions. However, they may ignore contextual importance, leading to rankings that do not align with practical needs. Therefore, improper or biased weight assignment can lead to skewed decision outcomes, resulting in ineffective or even harmful real-world implementations, especially in critical domains such as healthcare, transportation, and environmental policy. Comprehensive weight methods combine multiple weighting techniques to capitalize on the strengths and compensate for the limitations of each method. The most common approaches include:

Linear combination of subjective and objective weights.
Data-driven machine learning integration.
Fuzzy logic-based aggregation.
Bayesian and evidential reasoning frameworks.

These combine subjective and objective approaches to balance expert judgment and data-driven insights. These hybrid weighting methods may provide a balanced approach, improving ranking reliability. However, they are computationally complex and may require more data.

6.5.1 Impact on Ranking Order

The choice of weighting method has a direct impact on the ranking order of alternatives. Here's how comprehensive weighting influences this process:

1. Reduction in Bias and Uncertainty: Subjective methods often rely on expert knowledge and opinion, which may introduce personal biases or inconsistencies. Objective methods, although data-driven, may not capture contextual nuances. Comprehensive methods mitigate these issues by synthesizing expert knowledge with empirical data, leading to more balanced rankings.
2. Enhanced Robustness and Consistency: Using multiple weighting schemes can increase the robustness of the decision model. In scenarios where slight changes in criteria weights lead to significant ranking shifts, comprehensive methods tend to stabilize the rankings by averaging or intelligently integrating multiple views.
3. Better Stakeholder Acceptance: Stakeholders from diverse backgrounds, for example, technical, managerial, societal often question decisions based solely on subjective or purely mathematical criteria. Comprehensive methods provide a traceable, justifiable rationale by incorporating varied perspectives, increasing the acceptability of the final ranking order.
4. Handling Conflicting Criteria: Many real-world decisions involve trade-offs between conflicting criteria, for example, cost versus quality, efficiency versus safety. Comprehensive weight methods help balance these conflicts more effectively by blending subjective preferences which may favor one criterion with objective realities which favor another, leading to a more equitable ranking order.

Comprehensive weight methods offer a pragmatic and effective solution to one of the most critical challenges in MADM the determination of accurate, unbiased, and context-sensitive criteria weights. Their impact on the ranking order of alternatives is profound, often resulting in more balanced, acceptable, and robust decisions. While they do introduce additional complexity, their benefits in terms of increased reliability, reduced bias, and stakeholder alignment make them indispensable in modern decision-making frameworks. For high-stakes or multi-stakeholder problems, comprehensive weighting is no longer just an option but a necessity to ensure fairness, accuracy, and transparency in the decision-making process.

6.6 Effects of Weighting and Normalization on Ranking

In MADM, normalization and weighting are two critical steps that significantly influence the ranking of alternatives and selection of most appropriate alternative selection. The normalization ensures comparability while weighting emphasizes criteria importance. Both directly shape the decision matrix, influencing the ranking outcome. The choice of techniques must align with the problem context, stakeholder preferences, and data characteristics [50].

6.6.1 Impact of Normalization on Ranking

Normalization transforms each decision criteria values into a dimensionless scale and in range typically [0,1] to ensure comparability among criteria measured in different units. That means normalization plays a crucial role in MADM by converting each decision criteria values to a common scale, allowing fair comparison across criteria with different units or ranges. The choice of normalization technique such as min–max, vector, or max normalization can significantly affect the relative ranking order of alternatives, especially when decision criteria values vary widely. Different methods may emphasize or suppress certain data patterns, leading to shifts in rankings. Inconsistent or inappropriate normalization can distort the decision matrix and misrepresent performance. Therefore, selecting an appropriate normalization method is vital to preserve data integrity and ensure accurate, unbiased MADM ranking outcomes [51–53]. The detailed description of the various normalization techniques has already been given in chapter two. Following may be some of the crucial impact of the normalization of the decision criteria on the ranking.

1. *Different Techniques Yield Different Rankings*

   The same dataset normalized using max–min versus vector methods can produce different rankings.

   Example: In TOPSIS, vector normalization tends to reduce the dominance effect of large-value attributes.
2. *Handling Benefit Versus Cost Criteria*

   Misclassification (treating a cost criterion as a benefit) can reverse rankings.
3. *Outlier Sensitivity*

   Max–min normalization is highly affected by extreme values, whereas z-score is more robust.

6.6.2 *Impact of Weighting on Ranking*

Weighting significantly influences the ranking outcomes in MADM by assigning importance levels to each decision criterion. The assigned weights determine how much each decision criteria contributes to the final decision in ranking and selection of appropriate alternative, often amplifying or diminishing the impact of specific decision criteria. Even small variations in weights can cause substantial changes in the overall ranking, especially when alternatives have similar performance scores. Subjective or objective weighting methods, such as AHP or entropy, can lead to different rank orders. Therefore, careful selection and sensitivity analysis of weights are essential to ensure the robustness, fairness, and reliability of MADM-based decisions. A critical challenge in MADM techniques lies in accurately assessing the relative significance of each decision criterion to establish a robust ranking of alternatives and select the optimal alternative. In some complex decision scenarios, conventional subjective techniques to weight decision criteria, which often depend on the decision-maker's expertise, knowledge, and experience, may prove unreliable [54].

The weight of the criteria for each decision criterion reflects the amount of information to assess the performance score of each alternative. For some complex decision problems, the subjective approach of assigning significance to each decision criterion is either difficult or unreliable due to the decision-maker's reliance on their knowledge, experience, and expertise. Therefore, the subjective weight method is not appropriate in some decision scenarios, as specific characteristics could exist independently of the decision-maker's knowledge, experience, expertise, perspective, and way of thinking. On the other hand, in the objective weighting method, decision criteria serve as the core component, functioning as decision-makers and sources of information. The objective weight method is a more straightforward method without the intervention of any decision-maker. In the objective weight method, each attribute is taken as an information source, and the weight of significance of each criterion reflects the amount of information contained in each of them. Finding the weight distribution of each decision criterion that has high variability and supports stable ranking is still an open research challenge. Although there are still very few studies to tackle the problem related to measuring the credibility of the weighting and ranking method [55]. Following may be some of the crucial impact of the weights of the decision criteria on the ranking.

6.6.2.1 Equal Weights Versus Unequal Weights

If all criteria are equally weighted, the ranking depends solely on normalization.

Unequal weights amplify the effect of certain criteria, potentially changing the top-ranked alternative.

Sensitivity to Weight Changes

Small changes in weights can alter rankings, especially when alternatives have close performance scores.

Robustness analysis (e.g., Monte Carlo simulation) helps assess weight sensitivity.

The combined effect of weighting and normalization determines the final ranking. In weighting dominance scenarios, a significantly higher weight for one decision criterion may result in normalization having a minimal effect. In scenario where normalization-induced biasness some normalization methods such as vector normalization reduce the impact of high-magnitude criteria, indirectly altering weight influence. Weighting allows trade-offs, while some normalization methods, such as lexicographic, enforce strict priority.

6.7 Comparative Analysis of MADM Methods

The MADM methods are crucial tools for evaluating alternatives across multiple conflicting decision criteria. SAW is the simplest, using linear aggregation, but it assumes compensatory trade-offs and is sensitive to normalization. MEW introduces a multiplicative utility approach suitable for nonlinear modeling but unable to tackle decision matrices with zero or negative values. WASPAS combines SAW and MEW through balance coefficient factor to increase robustness, yet adds complexity. TOPSIS ranks alternatives based on their Euclidian distance from best and worst solutions, offering clear compromise decisions, though it is sensitive to the normalization method and does not account for inter-criteria correlation. The TOPSIS methods also face the problem of ranking abnormalities. EDAS evaluates alternatives based on their distance from the average, providing a realistic benchmark, but ignores extreme preferences. VIKOR emphasizes a compromise solution between utility and regret, making it ideal for multi-stakeholder environments, though it is sensitive to parameter settings. PROMETHEE uses pairwise comparisons with preference functions, supporting nuanced decision-making but requires expert knowledge to interpret flows effectively. GRA excels in uncertain or incomplete datasets through similarity measurement to an ideal reference, but lacks explicit trade-off modeling. Each method varies in dependence on normalization and weights, with distinct strengths and limitations, making method selection context-dependent based on data characteristics and decision-maker preferences.

There is no one-fits-all MADM method for all the problems. Different ranking methods have different weaknesses and strengths. Various ranking methods have different impacts on ranking and appropriate alternative selection of different weighting and normalization methods. A comparative analysis of the MADM methods discussed, such as SAW, MEW, WASPAS, TOPSIS, EDAS, VIKOR, PROMETHEE, and GRA, based on core principles, steps, normalization and weighting dependencies, strengths, weaknesses, limitations, challenges, and other key aspects, is summarized in Table 6.1.

Table 6.1 Summary of various MADM methods on different aspects

Method	Core principle	Normalization dependency	Weight dependency	Key strength	Major weakness	Best use case
SAW	Weighted sum	High	High	Simple	Linearity assumption	Basic evaluations
MEW	Multiplicative utility	High	High	Non-linear modeling	Not intuitive	Utility-based decisions
WASPAS	Hybrid of SAW and MEW	High	High	Robust	Complex	When robustness needed
TOPSIS	Ideal solution distance	High	High	Clear compromise	May mislead due to distances	Geometric compromise
EDAS	Distance from average	Medium	Medium	Realistic average ref	Ignores extremes	Moderate uncertainty
VIKOR	Compromise solution	High	High	Conflict resolution	Sensitive to v	Multi-stakeholder settings
PROMETHEE	Pairwise outranking	Medium	High	Rich preference modeling	Complex interpretation	DM preference-based
GRA	Grey relational similarity	High	Medium	Good under uncertainty	Weak trade-off modeling	Incomplete data problems

6.8 MATLAB Implementation

This section presents MATLAB implementations for each ranking method discussed earlier, using a fixed decision matrix and predefined criteria weight vectors. Readers can adapt the code to suit their specific applications by modifying the input matrix and weights as needed. These implementations serve as templates for applying MADM techniques in various real-world decision-making scenarios.

6.8.1 SAW Ranking Method

```
% Decision matrix (alternatives x criteria)
X = [4.5, 90, 8.1, 150, 5.5;
    3.8, 85, 7.9, 160, 6.0;
    4.2, 92, 8.3, 140, 5.0;
    4.0, 88, 8.0, 155, 5.8];
% This have this fixed decision criteria weight such that decision criteria weights (sum to 1)
weights = [0.25, 0.20, 0.25, 0.15, 0.15];
% Criteria types (1 = benefit, 0 = cost)
criteria_types = [1, 1, 1, 0, 0];
% Alternative names
alternatives = {'A1', 'A2', 'A3', 'A4'};
[m, n] = size(X);
normalized_X = zeros(m, n);
% Normalization
  for j = 1:n
     if criteria_types(j) == 1 % benefit
        normalized_X(:,j) = X(:,j) / max(X(:,j));
     else % cost
        normalized_X(:,j) = min(X(:,j)) ./ X(:,j);
     end
  end
 % Calculate scores
  scores = normalized_X * weights';
 % Ranking
  [~, ranking] = sort(scores, 'descend')
```

6.8.2 MEW Ranking Method

```
% Decision matrix (alternatives x criteria)
X = [4.5, 90, 8.1, 150, 5.5;
    3.8, 85, 7.9, 160, 6.0;
    4.2, 92, 8.3, 140, 5.0;
    4.0, 88, 8.0, 155, 5.8];
% This have this fixed decision criteria weight such that decison criteria weights (sum to 1)
weights = [0.25, 0.20, 0.25, 0.15, 0.15];
% Criteria types (1 = benefit, 0 = cost)
criteria_types = [1, 1, 1, 0, 0];
% Alternative names
alternatives = {'A1', 'A2', 'A3', 'A4'};
[m, n] = size(X);
normalized_X = zeros(m, n);
% Normalization
  for j = 1:n
     if criteria_types(j) == 1 % benefit
        normalized_X(:,j) = X(:,j) / max(X(:,j));
     else % cost
        normalized_X(:,j) = min(X(:,j)) ./ X(:,j);
     end
  end
% Calculate scores (product of normalized values raised to power of weights)
  scores = prod(normalized_X .^ weights, 2);
% Ranking
  [~, ranking] = sort(scores, 'descend')
```

6.8.3 WASPAS Ranking Method

```
% Decision matrix (alternatives x criteria)
X = [4.5, 90, 8.1, 150, 5.5;
    3.8, 85, 7.9, 160, 6.0;
    4.2, 92, 8.3, 140, 5.0;
    4.0, 88, 8.0, 155, 5.8];
% This have this fixed decision criteria weight such that decison criteria weights (sum to 1)
weights = [0.25, 0.20, 0.25, 0.15, 0.15];
% Criteria types (1 = benefit, 0 = cost)
criteria_types = [1, 1, 1, 0, 0];
% Alternative names
alternatives = {'A1', 'A2', 'A3', 'A4'};
[m, n] = size(X);
normalized_X = zeros(m, n);
lambda = 0.5; % It is the balance coefficent for involment of the two methods and default value
takem 0.5
% Normalization
  for j = 1:n
    if criteria_types(j) == 1 % benefit
      normalized_X(:,j) = X(:,j) / max(X(:,j));
    else % cost
      normalized_X(:,j) = min(X(:,j)) ./ X(:,j);
    end
  end
% SAW Method
  saw_part = normalized_X * weights';
% MEW Method
  mew_part = prod(normalized_X .^ weights, 2);
% WASPAS score which is integration of both SAW and MEW using balence
  scores = lambda * saw_part + (1 - lambda) * mew_part;
% Ranking
  [~, ranking] = sort(scores, 'descend')
```

6.8.4 TOPSIS Ranking Method

```
% Decision matrix (alternatives x criteria)
X = [4.5, 90, 8.1, 150, 5.5;
    3.8, 85, 7.9, 160, 6.0;
    4.2, 92, 8.3, 140, 5.0;
    4.0, 88, 8.0, 155, 5.8];
% This have this fixed decision criteria weight such that decison criteria weights (sum to 1)
weights = [0.25, 0.20, 0.25, 0.15, 0.15];
% Criteria types (1 = benefit, 0 = cost)
criteria_types = [1, 1, 1, 0, 0];
% Alternative names
alternatives = {'A1', 'A2', 'A3', 'A4'};
[m, n] = size(X);
normalized_X = zeros(m, n);
% Vector normalization
  for j = 1:n
     normalized_X(:,j) = X(:,j) / sqrt(sum(X(:,j).^2));
  end
% Weighted normalized decision matrix
  weighted_X = normalized_X .* weights;
% Determine ideal and negative-ideal solutions
  ideal = max(weighted_X);
  ideal(criteria_types == 0) = min(weighted_X(:, criteria_types == 0));
  neg_ideal = min(weighted_X);
  neg_ideal(criteria_types == 0) = max(weighted_X(:, criteria_types == 0));
% Calculate Euclidean distances
  d_plus = sqrt(sum((weighted_X - ideal).^2, 2));
  d_minus = sqrt(sum((weighted_X - neg_ideal).^2, 2));
% Calculate Performance scores
  scores = d_minus ./ (d_plus + d_minus);
% Ranking
  [~, ranking] = sort(scores, 'descend')
```

6.8.5 EDAS Ranking Method

```
% Decision matrix (alternatives x criteria)
X = [4.5, 90, 8.1, 150, 5.5;
     3.8, 85, 7.9, 160, 6.0;
     4.2, 92, 8.3, 140, 5.0;
     4.0, 88, 8.0, 155, 5.8];
  % This have this fixed decision criteria weight such that decison criteria weights (sum to 1)
  weights = [0.25, 0.20, 0.25, 0.15, 0.15];
  % Criteria types (1 = benefit, 0 = cost)
  criteria_types = [1, 1, 1, 0, 0];
  % Alternative names
  alternatives = {'A1', 'A2', 'A3', 'A4'};
  [m, n] = size(X);
  normalized_X = zeros(m, n);
  % Calculate average solution
     avj = mean(X, 1);
  % Calculate PDA and NDA
     PDA = zeros(m, n);
     NDA = zeros(m, n);
     for j = 1:n
        if criteria_types(j) == 1 % benefit
           PDA(:,j) = max(0, (X(:,j) - avj(j))) / avj(j);
           NDA(:,j) = max(0, (avj(j) - X(:,j))) / avj(j);
        else % cost
           PDA(:,j) = max(0, (avj(j) - X(:,j))) / avj(j);
           NDA(:,j) = max(0, (X(:,j) - avj(j))) / avj(j);
        end
     end
  % Calculate SP and SN
     SP = sum(PDA .* weights, 2);
     SN = sum(NDA .* weights, 2);
  % Normalize SP and SN
     SP_norm = SP / max(SP);
     SN_norm = 1 - (SN / max(SN));
  % Calculate appraisal score
     scores = 0.5 * (SP_norm + SN_norm);
   % Ranking
     [~, ranking] = sort(scores, 'descend')
```

6.8.6 VIKOR Ranking Method

```
% Decision matrix (alternatives x criteria)
X = [4.5, 90, 8.1, 150, 5.5;
    3.8, 85, 7.9, 160, 6.0;
    4.2, 92, 8.3, 140, 5.0;
    4.0, 88, 8.0, 155, 5.8];
% This have this fixed decision criteria weight such that decison criteria weights (sum to 1)
weights = [0.25, 0.20, 0.25, 0.15, 0.15];
% Criteria types (1 = benefit, 0 = cost)
criteria_types = [1, 1, 1, 0, 0];
% Alternative names
alternatives = {'A1', 'A2', 'A3', 'A4'};
[m, n] = size(X);
normalized_X = zeros(m, n);
v = 0.5; % default value
% Determine best and worst values
  f_plus = max(X);
  f_plus(criteria_types == 0) = min(X(:, criteria_types == 0));
  f_minus = min(X);
  f_minus(criteria_types == 0) = max(X(:, criteria_types == 0));
% Calculate S and R values
  S = zeros(m, 1);
  R = zeros(m, 1);
  for i = 1:m
    S(i) = sum(weights .* (f_plus - X(i,:)) ./ (f_plus - f_minus));
    R(i) = max(weights .* (f_plus - X(i,:)) ./ (f_plus - f_minus));
  end
% Calculate Q values
  S_plus = min(S); S_minus = max(S);
  R_plus = min(R); R_minus = max(R);
  Q = v * (S - S_plus) / (S_minus - S_plus) + ...
    (1 - v) * (R - R_plus) / (R_minus - R_plus);
  scores = Q;
% Ranking (lower Q is better)
  [~, ranking] = sort(scores, 'ascend')
```

6.8.7 PROMETHEE II Ranking Method

```
% Decision matrix (alternatives x criteria)
X = [4.5, 90, 8.1, 150, 5.5;
    3.8, 85, 7.9, 160, 6.0;
    4.2, 92, 8.3, 140, 5.0;
    4.0, 88, 8.0, 155, 5.8];
% This have this fixed decision criteria weight such that decison criteria weights (sum to 1)
weights = [0.25, 0.20, 0.25, 0.15, 0.15];
% Criteria types (1 = benefit, 0 = cost)
criteria_types = [1, 1, 1, 0, 0];
% Alternative names
alternatives = {'A1', 'A2', 'A3', 'A4'};
[m, n] = size(X);
normalized_X = zeros(m, n);
preference_function = 'usual'; % There are preference function such as quasi,usual,linear and
default
  % Initialize preference matrices
    P = zeros(m, m, n);
  % Calculate preference degrees for each criterion
    for j = 1:n
      for a = 1:m
        for b = 1:m
          d = X(a,j) - X(b,j);
          if criteria_types(j) == 0 % cost criterion
            d = -d;
          end
          % Apply preference function
          switch preference_function
            case 'usual'
              if d > 0
                P(a,b,j) = 1;
              else
                P(a,b,j) = 0;
              end
            case 'quasi'
              q = 0.1 * (max(X(:,j)) - min(X(:,j)));
              if d <= 0
                P(a,b,j) = 0;
              elseif d > q
                P(a,b,j) = 1;
              else
                P(a,b,j) = d / q;
              end
            case 'linear'
              p = 0.5 * (max(X(:,j)) - min(X(:,j)));
```

```
                if d <= 0
                    P(a,b,j) = 0;
                elseif d >= p
                    P(a,b,j) = 1;
                else
                    P(a,b,j) = d / p;
                end
            otherwise
                error('Unknown preference function');
        end
      end
    end
  end
% Aggregate preference indices
  pi = zeros(m, m);
  for a = 1:m
     for b = 1:m
        pi(a,b) = sum(weights .* squeeze(P(a,b,:)));
     end
  end
% Calculate positive and negative outranking flows
  phi_plus = sum(pi, 2) / (m - 1);
  phi_minus = sum(pi, 1)' / (m - 1);
% Calculate net flows
  net_flows = phi_plus - phi_minus;
% Ranking
  [~, ranking] = sort(net_flows, 'descend')
```

6.8.8 GRA Ranking Method

```
% Decision matrix (alternatives x criteria)
X = [4.5, 90, 8.1, 150, 5.5;
    3.8, 85, 7.9, 160, 6.0;
    4.2, 92, 8.3, 140, 5.0;
    4.0, 88, 8.0, 155, 5.8];
% This have this fixed decision criteria weight such that decison criteria weights (sum to 1)
weights = [0.25, 0.20, 0.25, 0.15, 0.15];
% Criteria types (1 = benefit, 0 = cost)
criteria_types = [1, 1, 1, 0, 0];
% Alternative names
alternatives = {'A1', 'A2', 'A3', 'A4'};
[m, n] = size(X);
normalized_X = zeros(m, n);
zeta = 0.5; % distinguishing coefficient
% Determine reference sequence
    ref_seq = max(X);
    ref_seq(criteria_types == 0) = min(X(:, criteria_types == 0));
% Normalize the decision matrix
    normalized_X = zeros(m, n);
    for j = 1:n
        if criteria_types(j) == 1 % benefit
            normalized_X(:,j) = X(:,j) / max(X(:,j));
        else % cost
            normalized_X(:,j) = min(X(:,j)) ./ X(:,j);
        end
    end
 % Normalize reference sequence
    normalized_ref = ones(1, n);
    % Calculate grey relational coefficient
    grey_coeff = zeros(m, n);
    for i = 1:m
        for j = 1:n
            delta = abs(normalized_ref(j) - normalized_X(i,j));
            grey_coeff(i,j) = (min(min(abs(normalized_ref - normalized_X))) + ...
                        zeta * max(max(abs(normalized_ref - normalized_X)))) / ...
                        (delta + zeta * max(max(abs(normalized_ref - normalized_X))));
        end
    end
    % Calculate grey relational grade
    scores = grey_coeff * weights';
    % Ranking
    [~, ranking] = sort(scores, 'descend')
```

6.9 Numerical Computation and Analysis

This section presents a numerical analysis of performance scores obtained from MATLAB implementations of the discussed ranking methods, using a fixed decision matrix and predefined criteria weights. The results help identify which ranking method may be more suitable for specific application types, supporting the effective use of MADM techniques in real-world decision-making scenarios. Consider a decision matrix with 4 alternatives and 5 associated decision criteria. Among these, three decision criteria are treated as a benefit criterion, while the remaining two are considered cost criteria. For simplicity in performance analysis, a fixed decision criteria weight vector has been used throughout the evaluation.

Table 6.2 presents the decision matrix, while Table 6.3 displays the weights assigned to each decision criterion, as determined by a specific weighting method. Tables 6.4 and 6.5 present the performance score of each alternative, and based on these performance scores, rankings of four alternatives (A1 to A4) using seven different MADM methods: SAW, MEW, WASPAS, TOPSIS, EDAS, VIKOR, and GRA by considering three benefits (C1, C2, C3) and two costs (C4, C5) as decision criteria. Each cell indicates the rank assigned to an alternative by a specific method, with 4 being the highest preference. Across most methods, A3 consistently receives the top rank (rank 4), suggesting it is the most favorable alternative. A1 is generally ranked third, while A2 consistently receives the lowest rank (rank 1) across all methods. A4 maintains a steady position at rank 2. This consistency indicates a strong agreement among the methods in evaluating the alternatives.

Table 6.2 Decision matrix along with type of decision criteria

Alternative/decision criteria	C1 (Benefit)	C2 (Benefit)	C3 (Benefit)	C4 (Cost)	C5 (Cost)
A1	4.5	90	8.1	150	5.5
A2	3.8	85	7.9	160	6.0
A3	4.2	92	8.3	140	5.0
A4	4.0	88	8.0	155	5.8

Table 6.3 Decision criteria weight vector

Weight/decision criteria	C1	C2	C3	C4	C5
Weight	0.25	0.20	0.25	0.15	0.15

Table 6.4 Performance score of each alternative from different ranking methods

	SAW	MEW	WASPAS	TOPSIS	EDAS	VIKOR	GRA
A1	0.965992	0.96547	0.965731	0.720386	0.82527	0.1885	0.757618
A2	0.890096	0.88885	0.889475	0	0	1	0.460152
A3	0.983333	0.9829	0.983117	0.716363	1	0	0.888889
A4	0.919285	0.91845	0.918866	0.275448	0.30103	0.62045	0.538838

Table 6.5 Ranking order of alternative from different ranking methods

	SAW	MEW	WASPAS	TOPSIS	EDAS	VIKOR	GRA
A1	3	3	3	1	3	3	3
A2	1	1	1	3	1	1	1
A3	4	4	4	4	4	4	4
A4	2	2	2	2	2	2	2

6.10 Summary

In this chapter, we have explored the core concepts and methodologies related to MADM. We begin by introducing the fundamentals of MADM and the necessity of weight determination methods to reflect the relative importance of different attributes in decision-making scenarios. In this chapter some of the vital MADM techniques are presented, followed by an examination of fuzzy MADM methods that address uncertainty and imprecision in input data. The chapter then analyzes how comprehensive weighting methods influence the ranking order of alternatives and discusses the effects of both weighting schemes and normalization techniques on final decision outcomes. A comparative analysis of different MADM methods highlights their advantages, limitations, and appropriate use cases. Additionally, the impact of proportion coefficient weight distributions on decision results is explored. The chapter concludes with practical aspects, including numerical computation and analysis of MADM problems, and demonstrates MATLAB-based implementation for effective modeling and solution development. A summary consolidates the key insights and methodologies discussed throughout the chapter.

References

1. E.K. Zavadskas, Z. Turskis, S. Kildienė, State of art surveys of overviews on MCDM/MADM methods. Technol. Econ. Dev. Econ. **20**(1), 165–179 (2014). https://doi.org/10.3846/20294913.2014.892037
2. G.O. Odu, Weighting methods for multi-criteria decision making technique. J. Appl. Sci. Environ. Manag. **23**(8), 1449 (2019). https://doi.org/10.4314/jasem.v23i8.7

3. M. Singh, M. Pant, A review of selected weighing methods in MCDM with a case study. Int. J. Syst. Assur. Eng. Manag. **12**(1), 126–144 (2021). https://doi.org/10.1007/s13198-020-01033-3
4. A. K. Yadav, K. Singh, The influence of different weighting methods on MADM ranking techniques and its impact on network selection for handover in HetNet, in *2023 International Conference on Artificial Intelligence and Smart Communication (AISC)* (2023), pp. 959–963. https://doi.org/10.1109/AISC56616.2023.10085305
5. A.K. Yadav, K. Singh, N.I. Arshad, M. Ferrara, A. Ahmadian, Y.I. Mesalam, MADM-based network selection and handover management in heterogeneous network: a comprehensive comparative analysis. Res. Eng. **21**, 101918 (2024)
6. A.K. Yadav et al., ECCANS: enhanced CRITIC-based context-aware network selection algorithm for 5G HetNet. Sustain. Energy Technol. Assess. **60**, 103477 (2023)
7. R. V. Rao, Introduction to multiple attribute decision-making (MADM) methods, in *Decision Making in the Manufacturing Environment*, in Springer Series in Advanced Manufacturing. London: Springer London (2007), pp. 27–41. https://doi.org/10.1007/978-1-84628-819-7_3
8. I. Kaliszewski, D. Podkopaev, Simple additive weighting—a metamodel for multiple criteria decision analysis methods. Expert Syst. Appl. **54**, 155–161 (2016). https://doi.org/10.1016/j.eswa.2016.01.042
9. V. Sihombing, et al., *Implementation of simple additive weighting algorithm in decision support system—IOPscience.*. Accessed 19 May 2025. https://doi.org/10.1088/1757-899X/1088/1/012014/meta
10. H. Taherdoost, *Analysis of Simple Additive Weighting Method (SAW) as a Multi- Attribute Decision-Making Technique: A Step-by-Step by Hamed Taherdoost* :: SSRN. https://papers.ssrn.com/sol3/papers.cfm?abstract_id=4626829. Accessed 19 May 2025
11. S. Sorooshian, N.A. Azizan, N. Ale Ebrahim, Weighted aggregated sum product assessment. Math. Model. Eng. Probl. (MMEP), **9**(4), 873–878 (2022)
12. M. Behzadian, S.K. Otaghsara, M. Yazdani, J. Ignatius, A state-of the-art survey of TOPSIS applications. Expert Syst. Appl. **39**(17), 13051–13069 (2012)
13. Y.-J. Lai, T.-Y. Liu, C.-L. Hwang, TOPSIS for MODM. Eur. J. Oper. Res. **76**(3), 486–500 (1994). https://doi.org/10.1016/0377-2217(94)90282-8
14. S. Chakraborty, TOPSIS and modified TOPSIS: a comparative analysis. Decis. Anal. J. **2**, 100021 (2022)
15. D.L. Olson, Comparison of weights in TOPSIS models. Math. Comput. Model. **40**(7–8), 721–727 (2004)
16. S. Chakraborty, *Evaluation Based on Distance From Average Solution (Edas) Method*, vol. 16. Accessed 19 May 2025. https://doi.org/10.1201/9781003377030-16/evaluation-based-distance-average-solution-edas-method-shankar-chakraborty-prasenjit-chatterjee-partha-protim-das
17. A. Alinezhad, J. Khalili, EDAS method, in *New Methods and Applications in Multiple Attribute Decision Making (MADM)*, vol. 277, in *International Series in Operations Research & Management Science*, vol. 277 (Cham: Springer International Publishing, 2019), pp. 149–155. https://doi.org/10.1007/978-3-030-15009-9_21
18. A.E. Torkayesh, M. Deveci, S. Karagoz, J. Antucheviciene, A state-of-the-art survey of evaluation based on distance from average solution (EDAS): developments and applications. Expert Syst. Appl. **221**, 119724 (2023). https://doi.org/10.1016/j.eswa.2023.119724
19. S. Opricovic, G.-H. Tzeng, Compromise solution by MCDM methods: A comparative analysis of VIKOR and TOPSIS. Eur. J. Oper. Res. **156**(2), 445–455 (2004). https://doi.org/10.1016/S0377-2217(03)00020-1
20. J.-P. Brans, B. Mareschal, Promethee methods, in *Multiple Criteria Decision Analysis: State of the Art Surveys*, vol. 78, in *International Series in Operations Research & Management Science*, vol. 78 (New York: Springer-Verlag, 2005), pp. 163–186. https://doi.org/10.1007/0-387-23081-5_5
21. J.-P. Brans, Y. De Smet, PROMETHEE methods, *Multiple criteria decision analysis: state of the art surveys* (2016), pp. 187–219

22. J.-P. Brans, P. Vincke, Note—a preference ranking organisation method: (The PROMETHEE method for multiple criteria decision-making). Manage. Sci. **31**(6), 647–656 (1985)
23. S.K. Amponsah, K.F. Darkwah, A. Inusah, Logistic preference function for preference ranking organization method for enrichment evaluation (PROMETHEE) decision analysis. Afr. J. Math. Comput. Sci. Res. **5**(6), 112–119 (2012)
24. J.P. Brans, P. Vincke, B. Mareschal, How to select and how to rank projects: the Promethee method. Eur. J. Oper. Res. **24**(2), 228–238 (1986). https://doi.org/10.1016/0377-2217(86)90044-5
25. J. Deng, J. Zhan, W.-Z. Wu, A ranking method with a preference relation based on the PROMETHEE method in incomplete multi-scale information systems. Inf. Sci. **608**, 1261–1282 (2022). https://doi.org/10.1016/j.ins.2022.07.033
26. V. Podvezko, A. Podviezko, Dependence of multi-criteria evaluation result on choice of preference functions and their parameters. Ukio Technologinis ir Ekonominis Vystymas **16**(1). Accessed 19 May 2025. https://doi.org/10.3846/tede.2010.09
27. N. Agrawal, Multi-criteria decision-making toward supplier selection: exploration of PROMETHEE II method. Benchmarking Int. J. **29**(7), 2122–2146 (2021). https://doi.org/10.1108/BIJ-02-2021-0071
28. M. Behzadian, R.B. Kazemzadeh, A. Albadvi, M. Aghdasi, PROMETHEE: a comprehensive literature review on methodologies and applications. Eur. J. Oper. Res. **200**(1), 198–215 (2010)
29. X. Qi, et al., PROMETHEE for prioritized criteria. Soft Comput (2019). Accessed 19 May 2025. https://doi.org/10.1007/s00500-019-04109-w
30. Y. Kuo, T. Yang, G.-W. Huang, The use of grey relational analysis in solving multiple attribute decision-making problems. Comput. Ind. Eng. **55**(1), 80–93 (2008)
31. M.N.D. Satria, Modification of grey relational analysis for dynamic criteria weighting in decision-making systems.IIUM Eng. J. https://journals.iium.edu.my/ejournal/index.php/iiumej/article/view/3494. Accessed 19 May 2025
32. S. Liu, Y. Yang, Y. Cao, N. Xie, A summary on the research of GRA models, in *Grey Systems: Theory and Application* (2013)
33. D. Julong, Introduction to grey system theory. J. Grey Syst. **1**(1), 1–24 (1989)
34. S. Liu, J.Y.L. Forrest, *Grey systems: theory and applications*. Springer Science & Business Media, (2010)
35. S. Liu, J. Forrest, Y. Yang, A brief introduction to grey systems theory, in *Proceedings of 2011 IEEE International Conference on Grey Systems and Intelligent Services* (2011), pp. 1–9. https://doi.org/10.1109/GSIS.2011.6044018
36. W. Pan, L. Jian, T. Liu, Grey system theory trends from 1991 to 2018: a bibliometric analysis and visualization. Scientometrics **121**(3), 1407–1434 (2019). https://doi.org/10.1007/s11192-019-03256-z
37. H. Wu, A. Hu, Y. Yang, Superiority analysis of energy and industrial structures based on a novel grey relational analysis model. Kybernetes (2025). Accessed 19 May 2025. https://doi.org/10.1108/k-11-2024-2977/full/html
38. C. Kahraman, Multi-criteria decision making methods and fuzzy sets, in *Fuzzy Multi-Criteria Decision Making*, vol. 16, ed. C. Kahraman, in springer optimization and its applications, vol. 16. (Boston, MA: Springer US, 2008), pp. 1–18. https://doi.org/10.1007/978-0-387-76813-7_1
39. S.K. Sahoo, S.S. Goswami, A comprehensive review of multiple criteria decision-making (MCDM) methods: advancements, applications, and future directions. Decis. Mak. Adv. **1**(1) (2023). https://doi.org/10.31181/dma1120237
40. T.-C. Chu, Y. Lin, An extension to fuzzy MCDM. Comput. Math. Appl. **57**(3), 445–454 (2009)
41. V. Sasirekha, M. Ilangkumaran, G. Sakthivel, A comparative analysis of fuzzy-based AHP derived MCDM methods to select the apt heterogeneous wireless network. Int. J. Inf. Decis. Sci. **8**(3), 227–253 (2016)
42. G.-S. Liang, Fuzzy MCDM based on ideal and anti-ideal concepts. Eur. J. Oper. Res. **112**(3), 682–691 (1999). https://doi.org/10.1016/S0377-2217(97)00410-4

43. A. Mardani, A. Jusoh, E.K. Zavadskas, Fuzzy multiple criteria decision-making techniques and applications—two decades review from 1994 to 2014. Expert Syst. Appl. **42**(8), 4126–4148 (2015). https://doi.org/10.1016/j.eswa.2015.01.003
44. G.R. Jahanshahloo, F.H. Lotfi, M. Izadikhah, Extension of the TOPSIS method for decision-making problems with fuzzy data. Appl. Math. Comput. **181**(2), 1544–1551 (2006)
45. L. Dymova, An approach to generalization of fuzzy TOPSIS method—ScienceDirect. Accessed 19 May 2025. https://www.sciencedirect.com/science/article/abs/pii/S0020025513001667.
46. K. Palczewski, W. Sałabun, The fuzzy TOPSIS applications in the last decade. Procedia Comput. Sci. **159**, 2294–2303 (2019). https://doi.org/10.1016/j.procs.2019.09.404
47. M. Braglia, et al. Fuzzy TOPSIS approach for failure mode, effects and criticality analysis. Qual. Reliab. Eng. Int. (Wiley Online Library, Braglia, 2003). Accessed 19 May 2025. https://doi.org/10.1002/qre.528
48. V. Pandey, H. Dincer, A review on TOPSIS method and its extensions for different applications with recent development. Soft Comput. Accessed 19 May 2025. https://doi.org/10.1007/s00500-023-09011-0
49. M.K. Srivastava, S. Gaur, A. Ohri, Analysing the effectiveness of MCDM and integrated weighting approaches in groundwater quality index development. Water Conserv Sci Eng **9**(2), 35 (2024). https://doi.org/10.1007/s41101-024-00267-7
50. N. Vafaei, R.A. Ribeiro, L.M. Camarinha-Matos, Selection of normalization technique for weighted average multi-criteria decision making, in *Technological Innovation for Resilient Systems*, ed. L.M. Camarinha-Matos, K.O. Adu-Kankam, M. Julashokri, (Cham: Springer International Publishing, 2018), pp. 43–52. https://doi.org/10.1007/978-3-319-78574-5_4
51. N. Vafaei, R.A. Ribeiro, L.M. Camarinha-Matos, Assessing normalization techniques for simple additive weighting method. Procedia Comput. Sci. **199**, 1229–1236 (2022). https://doi.org/10.1016/j.procs.2022.01.156
52. N. Petrović, V. Jovanović, M. Petrović, B. Nikolić, J. Mihajlović, Comparative investigation of normalization techniques and their influence on MCDM ranking—a case study. Spectr. Mech. Eng. Oper. Res. **2**(1) (2025). https://doi.org/10.31181/smeor21202542
53. S.T. Mhlanga, M. Lall, Influence of normalization techniques on multi-criteria decision-making methods. J. Phys. Conf. Ser. 012076 (IOP Publishing, 2022). Accessed 09 Apr 2025. https://doi.org/10.1088/1742-6596/2224/1/012076/meta
54. R. Kumar, D. Pamucar, A comprehensive and systematic review of multi-criteria decision-making (MCDM) methods to solve decision-making problems: two decades from 2004 to 2024. Spectr. Decis. Mak. Appl. **2**(1), 178–197 (2025)
55. C.-L. Hwang, A.S.M. Masud, *Multiple objective decision making—methods and applications: a state-of-the-art survey*, vol. 164 (Springer Science & Business Media, 2012). https://books.google.co.in/books?hl=en&lr=&id=M0noCAAAQBAJ&oi=fnd&pg=PA12&dq=Survey+of+objective+weight+methods+of+MADM&ots=0_QuTG3OEq&sig=9o2EEiKpZGQl85PPihpLv8iY42Q. Accessed 11 Dec 2023

Chapter 7
Challenges in MADM Models

Abstract MCDM methodologies face significant hurdles in real-world application due to several factors. Handling multiple, often conflicting criteria makes finding a truly optimal solution difficult, especially with more complexity. Establishing accurate criteria weights is prone to subjectivity and can drastically alter outcomes, becoming contentious in group settings with differing expert opinions and often lacking precise quantitative information. Dealing with hybrid and uncertain data, requiring the integration of quantitative and qualitative assessments and the management of vagueness through methods like fuzzy set theory, presents another major challenge. Methodological limitations, such as overlooking criteria interdependencies and the static nature of many techniques in dynamic environments, further complicate matters. Practical concerns include the high computational demands of some methods and the lack of transparency in single aggregated scores, hindering stakeholder trust. Finally, existing models struggle with subjectivity in expert input, oversimplification through single linguistic terms, and the inability to effectively process complex, uncertain, and interdependent data, particularly in group decision-making where critical details or varying expert importance may be overlooked.

Keywords Uncertainty · Expert bias · Conflicting criteria · Criteria correlation · Sensitivity · Ranking abnormality · Criteria interdependence · Scalability problems

7.1 Challenges in Handling Multiple Criteria

The application of multi-criteria decision making (MCDM) methodologies, while offering structured approaches to complex choices, is inherently accompanied by several significant challenges. These stem from both the intricate nature of real-world decision problems and the specific characteristics of the methods themselves. At its foundation, the very necessity for MCDM arises from the common scenario where decision-makers must evaluate multiple alternatives based on various, often conflicting, criteria [1]. It is frequently the case that no single alternative excels across all dimensions, making the identification of a truly optimal solution elusive [2]. This

A. K. Yadav et al., *Multi-attribute Decision-Making: Bridging Theory and Practice*, Transactions on Computer Systems and Networks,
https://doi.org/10.1007/978-981-95-4782-1_7

inherent complexity tends to escalate proportionally with the number of criteria and alternatives under consideration [3], demanding increasingly sophisticated analytical approaches.

Perhaps one of the most critical and persistently difficult tasks within the MCDM process is the establishment of suitable weights to represent the relative importance of each evaluation criterion. The process is fraught with potential pitfalls. Relying on arbitrary weights derived from purely subjective judgments can significantly compromise the accuracy and reliability of the final decision [3]. Decision-makers' (DMs) subjective assessments regarding criteria importance inevitably translate into subjective weights, which can directly impact the validity of the outcome; even subtle shifts in these judgments have the potential to drastically alter the final ranking of alternatives [4]. This challenge is often magnified in group decision contexts, where multiple experts may possess conflicting perspectives on criteria importance, rendering the process of achieving consensus on weights both difficult and potentially controversial. Furthermore, DMs may genuinely struggle to provide precise quantitative information about attribute weights, and in some situations, reliable information regarding [5] the relative importance of criteria might be partially or even completely unknown.

Another major hurdle involves effectively dealing with the hybrid and uncertain nature of data common in practical decision problems. Evaluations often require integrating both quantitative metrics and qualitative assessments. These criteria frequently employ different measurement units, posing significant difficulties for consistent assessment and subsequent aggregation [6]. Moreover, real-world scenarios are typically characterized by inherent vagueness, ambiguity, and complexity, leading to uncertainty in the evaluation data provided. Evaluators often utilize linguistic variables (such as "high," "medium," and "low") to convey their subjective judgments, necessitating the use of frameworks like fuzzy set theory and the integration of fuzzy numbers to adequately capture and process this valuable, albeit imprecise, information [3].

Beyond weighting and data complexities, methodological limitations also warrant attention. A crucial aspect often overlooked by traditional MCDM methods is the potential interdependencies between criteria; attributes are not always independent and can exhibit relationships of complementarity or redundancy, which, if ignored, can skew results [7]. The dynamic nature of many decision environments presents another challenge, as the importance or relevance of criteria may fluctuate over time. Conventional MCDM techniques, often static in their formulation, may fail to adequately consider the risk of unforeseen events that could necessitate a re-evaluation of criteria weightings. From a practical perspective, the computational demands of some MCDM methods can become substantial, particularly when scaling up to handle a large number of alternatives, criteria, or participating experts. Ensuring clarity and fostering trust in the results is also vital; methods that produce a single aggregated score for each alternative can sometimes lack transparency, making it difficult for stakeholders, especially those unfamiliar with the underlying calculations, to understand how the final ranking was derived and to fully trust the recommendation [3]. Finally, modern decision-making contexts demand flexibility, requiring methods capable of accommodating various assessment types including quantitative, qualitative [8], interval-valued, and mixed formats a capability not uniformly present across all existing MCDM approaches. Navigating these multifaceted challenges is essential for the effective and responsible application of MCDM techniques.

7.2 Aggregation Methods for Multiple Criteria

When dealing with decision-making under multiple objectives, the crucial step of synthesizing information from various criteria to arrive at a unified evaluation or ranking of alternatives is achieved through aggregation methods as shown in Fig. 7.1. These techniques represent a diverse toolkit, each with its own underlying philosophy and mathematical framework for combining performance data. We can broadly categorize these methods based on their primary source of input: those relying on expert judgment, those driven by objective data, and those that ingeniously blend both perspectives [5].

7.2.1 Subjective Aggregation Methods

Harnessing expert wisdom subjective aggregation methods place the insights and experience of experts at the forefront of the decision-making process. These approaches recognize that in many complex scenarios; human judgment carries invaluable information that might not be readily captured by purely quantitative data.

1. Pioneered by Thomas L. Saaty, the analytic hierarchy process (AHP) stands as a cornerstone of subjective MCDM. It provides a structured framework for tackling intricate problems with hierarchical structures, breaking them down into

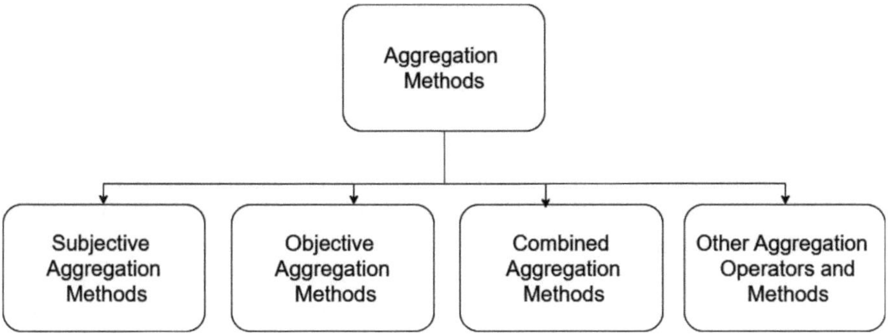

Fig. 7.1 Aggregation methods for multiple criteria

a series of levels encompassing the overall goal, criteria, sub-criteria (if applicable), and the alternatives. The core of AHP lies in its method of pairwise comparisons. Experts are asked to compare criteria against each other and alternatives against each criterion, expressing the relative importance or preference on a predefined scale. These qualitative judgments are then translated into quantitative values, forming individual comparison matrices. To derive a collective view, these individual matrices are aggregated using methods like the geometric mean method (which is sensitive to individual extreme judgments and provides a balanced aggregation) or the weighted arithmetic mean method (where each expert's opinion can be weighted based on their perceived expertise) [5]. A crucial aspect of AHP is the calculation of consistency ratios. These ratios assess the degree of logical consistency in the experts' pairwise judgments, ensuring the reliability and validity of the derived weights. A high inconsistency ratio signals potential issues in the comparisons, prompting a re-evaluation by the experts.

2. Recognizing that human reasoning is often imbued with vagueness and imprecision, the fuzzy analytic hierarchy process (FAHP) extends the traditional AHP by incorporating the principles of fuzzy logic. Instead of crisp numerical values, FAHP utilizes triangular fuzzy numbers (TFNs) to represent the subjective judgments of experts. These fuzzy numbers, defined by a lower bound, a most likely value, and an upper bound, allow for the expression of uncertainty and ambiguity in the pairwise comparisons. Similar to AHP, fuzzy pairwise comparison matrices are constructed. The aggregation of these fuzzy matrices typically employs the fuzzy geometric mean method, which is adapted to handle fuzzy numbers. Subsequently [1], fuzzy weights for the criteria and alternatives are calculated. These fuzzy weights are then often defuzzified into crisp values to facilitate the final ranking and decision-making. FAHP provides a more nuanced and realistic representation of expert opinions, particularly in situations characterized by incomplete information or inherent subjectivity.
3. In contrast to the detailed pairwise comparisons required by AHP and FAHP, the ordinal priority approach (OPA) offers a simpler and less cognitively demanding method for eliciting expert preferences. OPA focuses solely on the priority order of the criteria as determined by the experts. Instead of assigning numerical values or making direct comparisons of relative importance, experts are simply asked to rank the criteria according to their significance. This method avoids the need for detailed numerical input or pairwise comparison matrices, making it particularly useful in situations where experts may have sufficient knowledge to establish a clear ranking but might struggle with assigning precise numerical values. OPA then employs a specific mathematical procedure to derive initial weights for both the decision-makers (if multiple experts are involved) and the criteria based on these ordinal rankings [1]. This approach is particularly advantageous when a large number of criteria must be considered or when experts face time constraints or limited information for detailed evaluations, as it enables them to concentrate on what they know best, namely, the relative importance of the criteria.

7.2.2 *Objective Aggregation Methods*

Data-driven decision-making objective aggregation methods take a different tack, aiming to eliminate personal bias by relying solely on the quantitative data available for the decision problem. These methods derive criteria weights and alternative evaluations directly from the dataset, based on mathematical or statistical properties of the information.

1. The Maximum Deviation Method is another objective weighting technique used in (MADM). This method determines the importance of each criterion based on the total deviation of its values across all the alternatives. The underlying principle is that a criterion that exhibits larger deviations in performance among the alternatives is considered more important because it contributes more significantly to differentiating between them. The method calculates the deviation of each alternative's performance on a given criterion from the average performance across all alternatives for that criterion. The total deviation for each criterion is then summed up, and these sums are normalized to obtain the objective weights [9]. Criteria with larger total deviations receive higher weights, reflecting their greater influence on the relative ranking of the alternatives.
2. The Cochran-Weiss-Shanteau Ratio (CWS Ratio) stands out as an objective method specifically designed to measure the expertise of decision-makers (DMs). Unlike methods that focus on criteria or alternatives, the CWS ratio analyzes the consistency and agreement among the judgments provided by different experts. It assesses the extent to which an individual expert's judgments align with the collective judgments of the group and how consistent their own judgments are across different comparisons [7]. A higher CWS ratio indicates a greater level of perceived expertise. While primarily used to weight the opinions of different DMs in group decision-making scenarios, the CWS ratio can also be seen as an objective way to value the input of individuals based on the statistical properties of their responses.
3. The concept of additivity consistency of fuzzy preference relations (FPR) can also be employed as an objective weighting method. Fuzzy preference relations are used to represent the degree of preference between pairs of alternatives, where the preferences are expressed using fuzzy numbers to account for uncertainty. Additivity consistency refers to a specific mathematical property that indicates a logical and coherent structure within these preference relations. When a fuzzy preference relation exhibits high additivity consistency, it suggests a more reliable and well-structured set of preferences. This property can be quantified and used to derive objective weights for the criteria or even the decision-makers themselves [10]. In some contexts, additivity consistency has been adopted as a potential substitute for the need for replication often associated with the CWS ratio, offering an alternative way to assess the quality and consistency of subjective judgments.
4. The correlation coefficient and standard deviation (CCSD) technique is a statistical method employed in extensions of the MULTIMOORA method to compute

the relationships between different ranking strategies (ratio system approach, reference point approach, and full multiplicative form) and subsequently assign unique weights to them. This approach recognizes that different aggregation strategies might yield varying rankings, and the degree to which these rankings correlate or deviate can provide valuable information about the robustness and reliability of the results.

7.2.3 Combined Aggregation Methods

Synergizing subjectivity and objectivity recognizing the inherent strengths and limitations of purely subjective and purely objective approaches, combined aggregation methods aim to harness the synergy between expert insights and empirical data. These methods seek to create a more comprehensive and nuanced decision-making process by integrating both qualitative judgments and quantitative information [5]. Figure 7.2 presents a general approach to the combined aggregation methods, outlining its key steps and structure. This method integrates various subjective and objective weighting techniques to derive more balanced and accurate results in multi-criteria decision-making or evaluation processes.

1. A common and effective strategy for combining subjective and objective information involves integrating the weights obtained from subjective methods like AHP or FAHP with the weights derived from the objective entropy method.

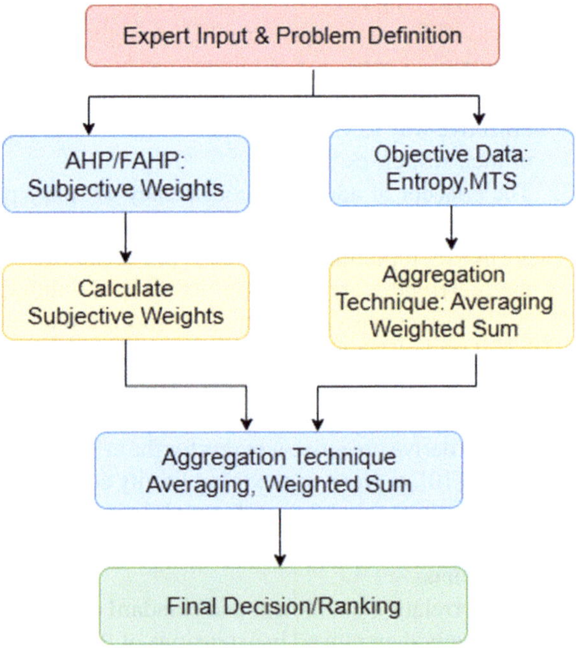

Fig. 7.2 Combined aggregation methods

In these hybrid approaches, both sets of weights are calculated independently. The subjective weights reflect the relative importance of criteria as perceived by experts, while the objective weights based on entropy reflect the information content and discriminatory power of the data itself.

These two sets of weights are then combined using various aggregation techniques [5], such as simple averaging or weighted averaging, to establish a set of combined weights for the criteria and sub-criteria. This integration allows the final decision to be influenced both by the expert's understanding of the problem domain and the objective characteristics of the data.

2. This sophisticated hybrid method combines the strengths of the Mahalanobis Taguchi System (MTS), a powerful data analysis technique for quality engineering, with the Choquet integral, a flexible aggregation function capable of handling the interdependencies between attributes. Typically, AHP might be used to determine the fuzzy integral of attributes, capturing the subjective preferences and relationships between them. Simultaneously, MTS can be employed to analyze objective data and identify patterns or deviations relevant to the decision problem. The Choquet integral is then used to aggregate the attribute values, taking into account the potential interactions and non-additivity between them, as defined by the fuzzy measures derived from AHP and the insights from MTS [11]. This approach allows for a highly nuanced aggregation that considers both subjective preferences, objective information, and the complex interplay between different criteria.
3. Integrated weighting methods represent a more direct approach to combining subjective and objective information by simultaneously considering both types of inputs to determine the importance of criteria. For example, subjective weights might be derived using methods like multiplicative AHP, while objective weights are calculated using techniques like the CWS ratio or FPR. These two sets of weights are then integrated using various mathematical operators (e.g., multiplication, averaging) [9] to arrive at a final set of criteria weights that reflect both expert judgment and data characteristics. This integrated approach provides a more holistic assessment of criteria importance, potentially leading to more robust and defensible decisions.

7.2.4 Other Aggregation Operators and Methods

A diverse toolkit beyond the broad categories of subjective, objective, and combined methods lies a diverse array of specific aggregation operators and MADM techniques, each designed to address particular aspects of the decision problem or handle specific types of information.

1. Fundamental operators like the weighted average (WA) operator are widely used for aggregating information based on the assigned weights of criteria and the performance scores of alternatives. This is particularly common in group decision-making, where individual expert preferences on alternatives are

weighted based on their expertise or the importance of the criteria. The geometric mean is another basic operator, often employed in the aggregation of pairwise comparisons in AHP to derive criteria weights or in FAHP to aggregate fuzzy judgments. The MAX–MIN method represents a more conservative approach, where the aggregated value is determined by the maximum of the minimum judgments, focusing on the worst-case scenario [6]. Simple additive weighting (SAW), also known as the weighted sum method, is a straightforward MADM technique that calculates the overall score of each alternative by summing the products of its performance rating on each criterion and the weight of that criterion.

2. With the increasing complexity of decision problems and the need to handle uncertain or imprecise information, a range of advanced aggregation operators have been developed. Power average (PA), power geometric (PG), and power-weighted geometric average (PWGA) operators are designed to aggregate data while considering the interrelationships between input values and potentially incorporating time factors, particularly in the context of interval-valued intuitionistic fuzzy sets (IVIFNs). Specific variations like SIIFPWGA and DIIFPWGA operators have been tailored for specific applications within this domain [3]. For handling linguistic uncertainty, aggregation operators for hesitant fuzzy linguistic term sets (HFLTSS) have been developed, often based on possibility distributions to aggregate the multiple possible linguistic terms provided by decision-makers.
3. To address decision problems where information is inherently vague or uncertain, extensions of MADM methods and specialized aggregation operators have been developed using fuzzy set theory and its extensions. MULTIMOORA (multi-objective optimization based on ratio analysis plus full multiplicative form) and its extended version using trapezoidal fuzzy neutrosophic numbers (TraFNNs) provide robust frameworks for evaluating alternatives under uncertainty. These methods often employ specific aggregation operators like TraFNNWA (trapezoidal fuzzy neutrosophic number weighted average) and TraFNNWG (trapezoidal fuzzy neutrosophic number weighted geometric) to combine the fuzzy neutrosophic information [4]. The interval-valued picture fuzzy geometric weighted Heronian average mean (IVPFGWHM) operator is a sophisticated tool designed for dynamic MADM problems with interrelated attributes and interval-valued picture fuzzy numbers, allowing for the representation of degrees of membership, non-membership, and hesitancy within intervals.

7.2.5 Sensitivity Analysis in Multi-Attribute Decision Making

In the intricate process of (MADM), where choices are made based on a multitude of criteria, the reliability and robustness of the final outcomes are paramount. Sensitivity analysis emerges as a critical process in this context, acting as a stress test for the decision model [5]. It systematically examines how variations in the input parameters and underlying conditions can influence the final ranking of alternatives or the selection of the most preferred option. By probing the stability of the results under different circumstances [5], sensitivity analysis provides decision-makers with a deeper understanding of the model's strengths and weaknesses, ultimately bolstering their confidence in the chosen course of action.

7.2.6 The Purpose of Sensitivity Analysis

Sensitivity analysis in MADM primarily aims to confirm the reliability and consistency of the chosen decision-making methods [5]. It assesses whether different approaches yield similar results and how changes in input data (like ratings or weights) impact the final rankings and potential errors [4, 8]. A key goal is to understand the method's resilience to uncertainty, especially variations in criteria weights, and to ensure the stability of the results, identifying the consistently top-ranked alternative even with changing conditions [3].

7.2.7 Methods of Conducting Sensitivity Analysis

Various techniques are employed to conduct sensitivity analysis in MADM, each designed to explore the impact of different types of variations on the decision outcomes.

1. Scenario-Based Analysis: This method involves constructing distinct scenarios by systematically varying multiple parameters or conditions simultaneously. Each scenario represents a plausible alternative state of the decision problem. By running the MADM model under each scenario, analysts can observe how the rankings of alternatives change in response to these combined variations [5]. This approach provides a holistic view of the model's sensitivity to complex and interacting changes.
2. Variation of Criteria Weights: Given that criteria weights often reflect subjective preferences and can significantly influence the final rankings, a common sensitivity analysis technique involves making incremental changes to these weights. By slightly adjusting the weights of different criteria (either individually or in combination) and observing the resulting shifts in the alternatives' rankings [8], decision-makers can understand how sensitive the outcomes are to changes in the perceived importance of different factors. In methods like AHP, this can be achieved by adjusting the initial pairwise comparisons while adhering to the underlying weighting logic.
3. Parameter Variation: Some MADM methods incorporate specific parameters within their mathematical formulations. In such cases, sensitivity analysis involves assigning different plausible values to these parameters and examining their effect on the aggregation results and the final rankings [12]. This helps to

understand the influence of these internal parameters on the decision outcome and to identify optimal or critical parameter settings.

4. Reversing Priorities: In group decision-making scenarios, where multiple experts provide their preferences, sensitivity analysis can involve reversing the priorities or preferences assigned by individual experts to the criteria. This helps to assess the impact of individual expert opinions and potential disagreements on the overall group decision [13]. By observing how the final rankings change when certain experts' priorities are inverted, the robustness of the consensus can be evaluated.
5. Considering Future Events: For decision problems in dynamic environments, sensitivity analysis can extend to considering different potential future events or states. Each future state might be associated with different criteria weightings or even the emergence of new criteria [7]. By analyzing the decision outcomes under these various future scenarios and considering the probabilities of their occurrence, decision-makers can assess the long-term robustness of their choices.

7.2.8 The Importance of Sensitivity Analysis

Sensitivity analysis is vital because it builds decision-makers' confidence by showing the stability of the chosen option across different scenarios. It also reveals how much the results are affected by changes in input data, highlighting critical factors that significantly drive the decision [5]. This understanding allows for a more thorough evaluation and ultimately leads to more informed and robust choices. Furthermore, it helps in comparing the reliability of different MADM methods under varying conditions [8], guiding the selection of the best approach, with hybrid and fuzzy methods often showing greater stability.

7.3 Ranking Abnormality

In the application of MCDM, the goal is to obtain a meaningful and reliable ranking of alternatives based on their performance across various criteria. However, the derived rankings are not always straightforward or predictable. The concept of ranking abnormality arises when these rankings exhibit unexpected behaviors, inconsistencies, or significant variations that warrant closer examination. Several factors inherent in the MCDM process can contribute to such anomalies. Figure 7.3 illustrates the factors influencing ranking abnormalities in a MACDM approach, highlighting key elements that contribute to inconsistencies or unexpected outcomes in multi-attribute decision-making processes.

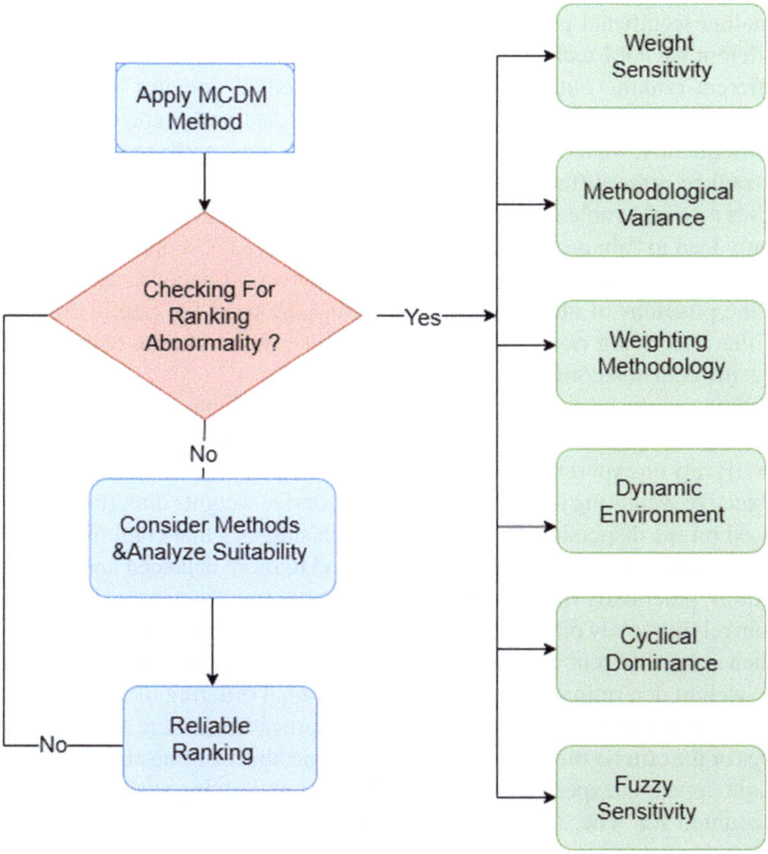

Fig. 7.3 Ranking abnormality

1. One primary source of ranking abnormality stems from the sensitivity of the final ranking to changes in the weights assigned to the criteria. As criteria weights often reflect the relative importance deemed by decision-makers (which can be subjective and subject to change), even seemingly minor adjustments to these weights can trigger substantial and unexpected shifts in the ranking order of alternatives [5]. Sensitivity analysis is a crucial tool for exploring this phenomenon, revealing how the stability of the rankings is affected by variations in the importance placed on different factors. For instance, changes in expert preferences can cascade through the weighting process, altering the significance of both high-level criteria and more specific attributes [13], ultimately leading to potentially surprising reorderings of the options. The observation of significant rank reversals with weight modifications underscores the importance of carefully considering the robustness of the derived rankings.

2. Another significant contributor to ranking abnormality is the fact that applying different MCDM techniques to the same decision problem can yield markedly different ranking outcomes. Each MCDM method operates based on its own unique set of principles, mathematical formulations, and aggregation procedures. Consequently, what one method identifies as the most preferred alternative might be ranked quite differently by another. This inherent methodological variance poses a considerable challenge in MCDM, as the choice of method can inadvertently lead to "abnormal" variations in the perceived best option [5]. Comparing the results obtained from various methods often reveals inconsistencies and swaps in the positions of alternatives, highlighting the need for careful consideration of the underlying assumptions and suitability of the chosen technique for the specific decision context.
3. Method employed for determining the criteria weights plays a pivotal role in shaping the final rankings. Subjective weighting methods, such as AHP and FAHP, rely on expert judgments, which, while valuable, can also introduce biases. Objective weighting methods, like entropy, derive weights directly from the data, based on the dispersion of values. Research suggests that combining subjective and objective weighting approaches can lead to more balanced and reliable evaluations, potentially mitigating some of the "abnormal" outcomes that might arise from relying solely on one type of method [5]. Discrepancies in rankings observed when using different weighting techniques underscore the significant impact that the weight determination process has on the final ordering of alternatives.
4. In dynamic multi-criteria decision-making processes, where the set of alternatives or the criteria themselves evolve over time, the resulting changes in rankings might appear unexpected or "abnormal" if the underlying shifts are not properly accounted for. The addition or removal of an alternative, or modifications to the performance scores of alternatives on the criteria, can logically lead to a new ranking. However, without a clear understanding of these changes and their implications, the resulting rank alterations might seem counterintuitive [8]. Sensitivity analysis in dynamic contexts helps to assess whether the observed ranking shifts are indeed logical consequences of the changes in the decision environment.
5. In certain preference evaluation scenarios, a situation known as circular reasoning can emerge. This occurs when a cycle of dominance exists, where alternative X is preferred to Y, Y is preferred to Z, and Z is preferred to X. Such a cyclical pattern violates the principle of transitivity and makes it impossible to establish a clear linear ranking. This phenomenon can be considered a form of "ranking abnormality" as it defies a straightforward ordering of preferences. Some decision theories [1], like dominance theory, might address this by ranking the involved alternatives at the same level, acknowledging the lack of strict preference.
6. In fuzzy MCDM methods, which are designed to handle the inherent vagueness and uncertainty in human judgments, the final ranking can exhibit sensitivity to minor variations in the membership and non-membership degrees assigned to alternatives with respect to the criteria. While sensitivity analysis in fuzzy contexts often reveals a degree of robustness, certain alternatives might be more susceptible to rank reversals even with slight changes in these fuzzy values

[4]. This potential for "abnormal" rank changes with seemingly insignificant input modifications highlights the importance of carefully considering the fuzzy representations and their impact on the final outcome.

7.4 MADM's Suitability and Complexity for Emerging Applications

MADM methods are essential for tackling the complexities of modern decision-making across various emerging applications. Their core strength lies in their structured approach to evaluating and prioritizing options based on multiple, often conflicting, criteria [5, 14]. This capability makes them highly relevant and widely applied in diverse cutting-edge fields, including supply chain optimization, cybersecurity, financial investments, resource management, technology selection, and sustainable energy development [15, 16], effectively addressing the intricate challenges of our rapidly evolving world.

1. Emerging applications often involve complex and uncertain data that combine both quantitative metrics and qualitative judgments. To manage this, (MADM) approaches are evolving with the use of fuzzy set theory, which effectively handles vagueness in expert opinions [11]. Advanced tools like interval-valued intuitionistic fuzzy sets (IVIFSs) and picture fuzzy numbers (PFNs) enhance the ability to capture deeper levels of uncertainty. Additionally, researchers are creating hybrid MADM (HMADM) methods to integrate and analyze diverse data formats [4]. These advancements ensure more comprehensive and adaptable decision-making frameworks.
2. A persistent challenge in MADM lies in the inherent subjectivity associated with the assignment of criteria weights by decision-makers. To mitigate the potential for bias and enhance the objectivity of the process [3], hybrid methodologies that strategically combine subjective weighting methods (such as the (AHP) and fuzzy AHP, which leverage expert opinions) with objective weighting methods (like the entropy method, which derives weights from data characteristics) are gaining prominence [5]. Additionally, the application of data-driven approaches, including techniques from the realm of machine learning, is being explored to derive more objective criteria weights and to discern the nature of each criterion (whether it represents a benefit to be maximized or a cost to be minimized).
3. Many emerging applications operate within dynamic environments where the available alternatives and the relevant criteria can change over time. Moreover, the attributes of these alternatives are often interdependent, meaning that the performance on one criterion can influence the performance on others [4]. To address these complexities, novel MADM approaches are being developed. These include methods that incorporate dynamic calculations to adapt to temporal changes and operators specifically designed to account for the interrelationships between data

points [8]. The stratified multicriteria decision-making (SMCDM) method exemplifies an approach that explicitly aims to integrate "likely future events" and their potential impact on criteria weights into the decision-making process [7].
4. Computational demands of different MADM methods can vary significantly. While some techniques, such as the ordinal priority approach (OPA), are designed for simplicity and computational efficiency, others, particularly hybrid or extended methods that grapple with complex data types and dynamic aspects, can be considerably more computationally intensive [2]. Researchers are actively engaged in efforts to enhance both the efficiency and the accuracy of these advanced MADM methods [7], striving for a balance between analytical rigor and practical applicability.
5. To handle the unique complexities of emerging applications, MADM is often combined with other techniques. For example, rough set theory and OPA enhance QFD, while DEA, fuzzy BWM, and simulation-based fuzzy TODIM aid wave energy site selection [10]. Evidence theory and fuzzy discrimination analysis are used for interactive MADM in emergency diagnostics [17]. This trend of integrating analytical tools provides more comprehensive and insightful solutions for modern challenge.

7.5 Future Trends in Multi-Attribute Decision-Making

The field of (MADM) is not static; it is a dynamic and evolving area of research driven by the need to address increasingly complex real-world problems. Several key trends and future directions indicate the exciting trajectory of MADM, focusing on enhancing its capabilities to handle new challenges and expand its applicability across diverse domains.

1. As decision-making scenarios grow in scale, involving a multitude of stakeholders and generating vast amounts of information, the limitations of traditional group decision-making (GDM) methods become apparent. A significant future trend in MADM is the development of methods specifically designed for large-scale group decision-,aking (LSGDM) [6]. This involves creating optimized approaches capable of efficiently managing and processing the extensive data generated in such contexts, ensuring that the collective wisdom of a large group can be effectively harnessed.
2. The reality of many decision problems is that they involve data that changes over time, often with interdependencies between different data points [18]. Recognizing this dynamic nature, a key future direction in MADM is the development of novel approaches that can effectively handle time-dependent data. This includes incorporating dynamic calculations, considering time preferences in the evaluation process [7], and utilizing methods like the (SMCDM) to account for potential future events and their impact on criteria weights, thus integrating aspects of risk and uncertainty related to time.

3. The inherent vagueness and uncertainty of real-world decision problems necessitate the use of tools that can effectively represent and process imprecise information. A continuing and significant trend in MADM is the integration of fuzzy set theory and linguistic representations [2]. This involves employing various advanced extensions of fuzzy sets, such as interval-valued intuitionistic fuzzy sets (IVIFS), picture fuzzy numbers (PFN), q-rung ortho pair fuzzy sets (q-ROFS), trapezoidal fuzzy neutrosophic sets, and bipolar complex fuzzy information [12, 17]. Crucially, the development of aggregation operators specifically tailored for these fuzzy environments is also a vital area of ongoing research [1, 16].
4. The synergistic potential of combining different MADM techniques and integrating them with other methodologies is a powerful driver of innovation in the field [13]. Hybridization of MADM methods is a growing trend, with examples including the integration of AHP with TOPSIS, the combination of the (MTS) with AHP and the Choquet integral, and the integration of rough set theory with the (OPA) [8, 10]. Future trends also point toward the integration of simulation-based optimization techniques with MCDM frameworks to create even more robust and versatile decision support systems.
5. To address the inherent subjectivity in traditional MADM and enhance the effectiveness of decision-making processes, there is a significant move toward incorporating data-driven methodologies and machine learning techniques. Future work aims to expand the applicability of these approaches to handle mixed data types [3], including fuzzy variables, promising a more objective and potentially more accurate way of deriving insights and supporting decisions.
6. Recognizing that the nuances of different application areas often require tailored solutions, there is an increasing focus on adapting and applying MADM methods to address challenges in specific domains [4]. Examples include supplier selection, healthcare, energy, IoT security, and logistics. The development of domain-specific taxonomies and frameworks, such as those for IoT security challenges [10, 15, 17], indicates a future trend toward more specialized and context-aware MADM applications.
7. Traditional MADM methods often assume that the attributes or criteria under consideration are independent. However, in many real-world scenarios, attributes are interconnected, and the performance on one can influence others. Therefore, future research is increasingly focusing on incorporating the interrelationships between attributes into MADM models [6]. Methods that utilize fuzzy measures and the Choquet integral, which can account for non-additive interactions, are particularly relevant in this context.
8. In group decision-making, achieving a satisfactory level of agreement among experts is crucial for the legitimacy and acceptance of the final decision. Future trends in GDM focus on developing methods that facilitate and effectively measure consensus among experts [17]. This includes developing consensus processes for various types of preference relations, such as probabilistic linguistic preference relations [19], and even leveraging sentiment analysis to detect cooperative groups and potentially enhance the consensus-building process.

7.6 Existing Model Limitations

In the evolving landscape of (MCDM), existing models face numerous challenges that limit their applicability and accuracy in complex, real-world scenarios [5]. One of the primary concerns is the subjectivity and uncertainty involved in expert-based models. Traditional techniques often rely heavily on subjective opinions, which are inherently ambiguous and may not fully capture the depth of expert knowledge [20]. For example, decision-makers might be constrained to choosing a single linguistic term, which oversimplifies their nuanced understanding.

Moreover, many models struggle to handle complex information effectively. Classical approaches such as crisp sets use binary membership functions, making them ill-equipped to deal with the vagueness and ambiguity present in real-life data [16]. Additionally, they lack the ability to process bipolar fuzzy information or deal with time-dependent and interdependent datasets [18]. Aggregating such intricate information—while accounting for interrelationships and weight influences remains a challenge for traditional aggregation operators.

Group decision-making introduces another layer of complexity. Some methods generalize alternative evaluations rather than basing them on specific characteristics, which leads to a loss of critical detail [6]. They may also fail to address the differing importance each expert assigns to the criteria, sometimes imposing a uniform weighting system that does not reflect individual perspectives [20]. Even advanced approaches, like those based on hesitant fuzzy linguistic term sets (HFLTS), may not fully overcome these limitations in group settings.

7.6.1 Several Method-Specific Limitations also Emerge upon Deeper Analysis

- TOPSIS (Technique for Order of Preference by Similarity to Ideal Solution), although popular, is sensitive to the choice of normalization technique. Its standard aggregation of distances from the Positive Ideal Solution (PIS) and Negative Ideal Solution (NIS) doesn't fully consider the relative significance of these separations [21]. Modified TOPSIS versions often suffer from inconsistencies in weight assignment, which can lead to different rankings.
- Analytic hierarchy process (AHP) and analytic network process (ANP) are well-known hierarchical methods, but both are prone to inconsistency, especially when the number of pairwise comparisons increases [20]. As a solution, fuzzy best–worst method (BWM) has been introduced [10] to reduce the number of comparisons while ensuring optimal consistency.
- Multi-attribute Utility Theory (MAUT) encounters difficulty when assessing decision-makers' risk behavior [3]. Its inability to effectively model interdependencies among criteria becomes particularly problematic in large-scale, multi-decision-maker problems.

- ELECTRE and TOPSIS are often not suitable for poorly structured problems. Their rigid frameworks lack the flexibility to incorporate hybrid assessments or extend to group decisions involving multiple experts [2]. They also require numerous external parameters, adding to the model complexity.
- Quality Function Deployment (QFD), though useful in product and process development, can become excessively complex when improved versions demand precise expert inputs and multiple aggregation techniques. In uncertain environments, traditional QFD shows limitations, and integrating multiple decision-making methods only adds to the time and financial cost.
- Stratified Multi-criteria Decision-Making (SMCDM) faces scalability issues. As the number of possible events increases, so does the computational burden [7], making it difficult to manage in expansive decision spaces.
- Subjective models, in general, pose risks when used in isolation under uncertainty [5], as they might lack the robustness needed for objective, well-rounded decision-making.

Furthermore, limitations in data availability and data type also affect the performance of existing MCDM methods [17, 22]. Many cannot handle decision matrices that contain both qualitative and quantitative data [14], making it challenging to build robust models in environments where objective, statistical data is scarce.

7.7 Future Research Directions

The field of (MCDM) is a vibrant and continuously evolving area. Building upon existing foundations, several exciting future research directions promise to expand the scope, enhance the capabilities, and broaden the applicability of MCDM methodologies. These future endeavors aim to address current limitations, leverage emerging technologies, and tackle increasingly complex real-world challenges.

1. A significant thrust of future research involves expanding the scope and applicability of existing MCDM methods. This includes exploring a wider range of criteria and alternatives in various problem contexts, such as supplier selection and supply chain performance evaluation [5, 20]. Furthermore, there is a clear direction toward applying proposed and existing MADM models to diverse domains and real-world problems to validate their effectiveness and identify their strengths and weaknesses across different scenarios [21]. This also encompasses investigating the applicability of specific MADM approaches to different types of data [10], including interval numbers and various forms of fuzzy numbers, to enhance their versatility.
2. The integration of MCDM with cutting-edge technologies presents a wealth of opportunities. A particularly promising avenue is the synergy between MCDM methods and artificial intelligence (AI) [5]. Future research is likely to explore how AI can be leveraged for predictive analysis, dynamic decision-making, and potentially automating aspects of the MCDM process, particularly in

dynamic environments like supplier selection. Moreover, there is an emphasis on combining existing MCDM tools and techniques in novel ways to create more robust and comprehensive frameworks, such as enhancing the SMCDM approach through integration [7, 20]. The application of advanced statistical techniques like structural equation modeling to analyze relationships within MCDM frameworks is also a potential future direction.

3. Handling increasingly complex and uncertain environments remains a core focus of future MCDM research. This involves the continued exploration and incorporation of various types of fuzzy sets and linguistic models to better capture the nuances of imprecise information [23]. This includes extending existing fuzzy MADM methods to accommodate more advanced fuzzy representations like intuitionistic fuzzy numbers and hesitant fuzzy information, as well as integrating methods like MULTIMOORA with sophisticated fuzzy set extensions. A crucial step forward is the development of methods capable of effectively handling decision matrices containing a mix of quantitative and qualitative (fuzzy) data [7]. Furthermore, refining dynamic calculations within dynamic MCDM approaches, particularly for time-critical situations like emergencies, is expected to yield even more effective decision support.
4. Improving the processes and outcomes of group decision-making (GDM), especially in Large-Scale Group Decision-Making (LSGDM) scenarios, is another key area of future research. This includes extending consensus-based methods to accommodate the increased volume of information and number of participants in LSGDM [6]. Further investigation into the strengths and weaknesses of different aggregation approaches for combining diverse expert opinions is also crucial. Additionally, incorporating more nuanced aspects like the ranking or expertise of individual experts [13, 24] within methods like enhanced QFD is a potential future development.
5. Continued efforts to refine and improve the existing arsenal of MCDM methods are essential. This includes exploring alternative defuzzification methods in fuzzy MCDM to enhance consistency and optimization [21]. For methods like TOPSIS, developing approaches for aggregated separation measures to streamline the process is a proposed direction [13]. Furthermore, the further development and application of newer methods like the rough ordinal priority approach (OPA-R) to a wider range of MCDA problems is anticipated.
6. Tailoring MCDM methodologies to the unique challenges of specific application areas is a growing trend [15]. Future research is expected to see a deeper dive into domains like IoT security, involving comprehensive literature reviews, expert case studies, and the development of specific guidelines [14]. In energy-related applications, this includes enriching the types of resources considered in allocation plans and exploring various relevant scenarios. For areas like regional safety assessment, continued evaluation of existing methods and the investigation of dynamic calculations are planned. Similarly [8, 13], further refinement and application of MCDM in frameworks like QFD for specific contexts are anticipated.

7. Underpinning the practical advancements in MCDM are ongoing theoretical developments. This includes further expanding the operational rules and related theories of advanced fuzzy number systems, such as interval-valued picture fuzzy numbers (IVPFN), including the development of new aggregation operators and the study of similarity and distance measures [12, 17]. Another potential theoretical advancement lies in embedding more sophisticated ways of accounting for attribute interactions, such as fuzzy discriminations, into aggregation operators like the Choquet integral.

In essence, the future of MCDM research is characterized by a drive toward greater sophistication, integration, adaptability, and domain-specific relevance. By addressing the limitations of current methods, leveraging emerging technologies, and delving into theoretical advancements [5], the field is poised to provide even more powerful and insightful tools for tackling the complex decision-making challenges of the future.

7.8 Summary

The application of multi-criteria decision making (MCDM) methodologies addresses complex decision problems involving multiple, often conflicting criteria, but it faces several key challenges. These include the difficulty of assigning appropriate weights to criteria, which often relies on subjective judgments that can vary widely and affect the final rankings. Group decision-making complicates this further due to differing expert opinions. Additionally, the hybrid nature of data—combining quantitative and qualitative information measured in different units introduces uncertainty and vagueness, often requiring fuzzy logic to handle linguistic evaluations. Methodological limitations also arise from ignoring interdependencies among criteria, the static nature of many methods that do not adapt to changing conditions, computational complexity with large datasets, and a lack of transparency that can reduce stakeholder trust. Aggregation methods, essential for synthesizing multiple criteria into unified evaluations, fall into three categories: subjective methods based on expert judgment (such as AHP, FAHP, and OPA), objective data-driven methods (including maximum deviation, Cochran Weiss-Shanteau ratio, and CCSD techniques), and combined approaches that integrate both subjective and objective inputs. Hybrid methods, like combining AHP with entropy weighting or using the Mahalanobis-Taguchi System alongside the Choquet integral, leverage the strengths of both perspectives to better capture uncertainty, interdependencies, and data patterns, thus enhancing decision robustness and reliability. Overall, effective MCDM requires balancing expert insights with empirical data while managing complexity, uncertainty, and methodological constraints to support transparent and trustworthy decisions.

References

1. I. Irvanizam, Z. Zulfan, P.F. Nasir, M. Marzuki, S. Rusdiana, N. Salwa, An extended multimoora based on trapezoidal fuzzy neutrosophic sets and objective weighting method in group decision-making. IEEE Access **10**, 47476–47498 (2022)
2. A.V. Demidovskij, Comparative analysis of madm approaches: electre, topsis and multi-level LDM methodology, in *2020 XXIII International Conference on Soft Computing and Measurements (SCM)* (2020), pp. 190–193
3. M. Torkjazi, A.K. Raz, Data-driven approach with machine learning to reduce subjectivity in multi-attribute decision making methods, in *2023 IEEE International Systems Conference (SysCon)* (2023), pp. 1–8
4. B. Chen, Y. Guo, X. Gao, Y. Wang, A novel multiattribute decision making approach: addressing the complexity of time dependent and interdependent data. IEEE Access **6**, 55838–55849 (2018)
5. I. Boukrouh, F. Tayalati, A. Azmani, A comprehensive framework for supplier selection: using subjective, objective, and hybrid multi-criteria decision-making techniques with sensitivity analysis. IEEE Access **12**, 145550–145569 (2024)
6. J.R. Trillo, F.J. Cabrerizo, M.J. Del Moral, J.A. Morente-Molinera, J.M. Tapia, E. Herrera-Viedma, A consensus-based multi-criteria group decision making method based on an aggregated operator customised by experts, in *2023 9th International Conference on Control, Decision and Information Technologies (CoDIT)* (2023), pp. 1302–1307
7. M.R. Asadabadi, The stratified multicriteria decision-making method: An extended version. IEEE Eng. Manag. Rev. **51**(2), 46–54 (2023)
8. J.P. Caylor, R.J. Hammell, Dynamic and fuzzy multicriteria decision-making for regional safety assessment ranking, in *2022 Joint 12th International Conference on Soft Computing and Intelligent Systems and 23rd International Symposium on Advanced Intelligent Systems (SCISISIS)* (2022), pp. 1–6
9. N.A. Masruroh, D.A. Rosalia, Multiple criteria group decision making with integrated objective and subjective weights for determining flagship industry ranking, in *2019 5th International Conference on Science and Technology (ICST)*, vol. 1 (2019), pp. 1–5
10. C.-N. Wang, N.-L. Nhieu, H.-P. Nguyen, J.W. Wang, Simulation-based optimization integrated multiple criteria decision-making framework for wave energy site selection: a case study of Australia. IEEE Access **9**, 167458–167476 (2021)
11. M.Z. Zhang, N.D. Yang, X.L. Zhu, A hybrid multiple attribute decision-making method based on mahalanobis-taguchi system and choquet integral, in *2021 IEEE International Conference on Industrial Engineering and Engineering Management (IEEM)* (2021), pp. 523–527
12. J.-P. Fan, H. Zhang, W. Mei-Qin, Dynamic multi-attribute decision-making based on interval-valued picture fuzzy geometric heronian mean operators. IEEE Access **10**, 12070–12083 (2022)
13. J.-L. Du, S.-F. Liu, S.A. Javed, M. Goh, Z.S. Chen, Enhancing quality function deployment through the integration of rough set and ordinal priority approach: a case study in electric vehicle manufacturing. IEEE Trans. Eng. Manag. **71**, 7541–7552 (2024)
14. R. Han, Q. Hu, Z. Guo, X. Quan, Z. Wu, R. Hu, Optimal allocation method of residential air-conditioners: trade-off solutions between economic costs and aggregation reliability. IEEE Open Access J. Power Energy, **9**, 131–142 (2022)
15. M.A. Akbar, A. Alsanad, S. Mahmood, A. Alothaim, A multicriteria decision making taxonomy of iot security challenging factors. IEEE Access **9**, 128841–128861 (2021)
16. Y. Song, T. Mahmood, U. Ur Rehman, Selection of software development methodology by employing a multi-criteria decisionmaking approach based on logarithmic bipolar complex fuzzy aggregation operators. IEEE Access **12**, 38163–38179 (2024)
17. G. Sirbiladze, J. Kacprzyk, B. Midodashvili, B. Ghvaberidze, J.A. Lozano, T. Davitashvili, Evidence theory and fuzzy discriminations in dependable and resilient interactive madm. Application in emergency psychiatric diagnostics, in *2023 13th International Conference on Dependable Systems, Services and Technologies (DESSERT)* (2023), pp. 1–6

18. F. Jahani, M. Mohammadi, A. Pourdaryaei, E. Shayesteh, M. Karimi, A novel multi-criteria decision-making framework in electrical utilities based on gray number approach. IEEE Access **10**, 121508–121519 (2022)
19. J.A. Morente-Molinera, I.J. Cabrerizo, S. Alonso, I.J. Pérez, E. Herrera-Viedma, Managing changes in alternatives and criteria during a dynamic multi-criteria group decision making process, in *2020 7th International Conference on Control, Decision and Information Technologies (CoDIT)*, vol. 1 (2020), pp 1–6
20. M. Divsalar, M. Ahmadi, Y. Nemati, A scor-based model to evaluate larg supply chain performance using a hybrid madm method. IEEE Trans. Eng. Manage. **69**(4), 1101–1120 (2022)
21. S. Abootalebi, A. Hadi-Vencheh, A. Jamshidi, Ranking the alternatives with a modified topsis method in multiple attribute decision making problems. IEEE Trans. Eng. Manage. **69**(5), 1800–1805 (2022)
22. S. Pandey, J. Han, N. Gurung, H. Chen, E.A. Paaso, Z. Li, A. Khodaei, Multi-criteria decision-making and robust optimization methodology for generator sizing of a microgrid. IEEE Access **9**, 142264–142275 (2021)
23. Y. Lin, W. Li, G. Liu, D. Huang, Method for triangular fuzzy multiple attribute decision making based on two-dimensional density operator method. J. Syst. Eng. Electron. **35**(1), 178–185 (2024)
24. S. Guo, Z. Qi, A fuzzy best-worst multi-criteria group decision making method. IEEE Access **9**, 118941–118952 (2021)

Chapter 8
Application of MADM

Abstract Applications of MADM models across key sectors such as sustainable development, smart cities, healthcare management, energy, agriculture, disaster management, climate change mitigation, and mobility management in 5G/6G implementation. MADM serves as a central decision-making framework, enabling the evaluation and prioritization of complex alternatives based on multiple criteria. For instance, in smart cities, MADM can optimize urban infrastructure projects by balancing cost, efficiency, and environmental impact, while in healthcare, it aids in resource allocation and policy formulation. Similarly, in agriculture, it supports precision farming decisions, and in disaster management, it enhances preparedness and response strategies. Figure 8.1 highlights how MADM bridges technology, sustainability, and governance, offering a structured approach to solving multifaceted challenges in diverse fields. This visualization underscores the model's versatility and its critical role in data-driven, multi-criteria decision-making processes across global and sector-specific contexts.

Keywords MADM · Decision support · Sustainable development · Healthcare decision-making · Energy management · IoT · Agricultural systems

8.1 Sustainable Development

The concept of sustainable development in the context of tourism and climate change emphasizes balancing economic growth, environmental protection, and social well-being to achieve long-term sustainability. Given the significant carbon emissions from tourism activities such as transportation, accommodation, and catering the industry must adopt low-carbon strategies to align with China's "dual carbon" goals (carbon peak by 2030 and carbon neutrality by 2060) [1]. Sustainable tourism development requires optimizing energy structures, promoting clean and renewable energy, and integrating green practices across urban planning and tourism management. Key measures include enhancing carbon sequestration through afforestation in scenic areas, implementing intelligent tourism systems to monitor and reduce

A. K. Yadav et al., *Multi-attribute Decision-Making: Bridging Theory and Practice*,
Transactions on Computer Systems and Networks,
https://doi.org/10.1007/978-981-95-4782-1_8

energy consumption, and fostering low-carbon travel behaviors among tourists. By addressing high-emission sectors and leveraging natural ecosystems for carbon absorption, the tourism industry can contribute to ecological civilization and sustainable urban development. This approach not only mitigates climate impacts but also ensures the long-term viability of tourist destinations, aligning with global sustainability frameworks and China's national policies on green transformation. The rapid pace of urbanization in China has led to impressive growth in urban infrastructure and building development. However, this expansion must be balanced with environmental preservation to ensure sustainable resource utilization. As society progresses and living standards improve, there is a growing demand not only for comfort but also for environmentally responsible development practices (Fig. 8.1).

Green building has emerged as a strategic response, aiming to harmonize economic growth with ecological balance [2]. It goes beyond individual structures and requires integration with broader urban planning. A sustainable green building approach necessitates interdisciplinary collaboration across architectural, structural, mechanical, and electrical domains. Standardizing technical design solutions for green buildings enhances designers' ability to address complex challenges holistically and supports the effective evaluation and implementation of sustainable practices. Moreover, these standardized solutions can guide developers through informed decision-making, enabling better control over project development and promoting long-term environmental and economic sustainability. As such, green building design plays a pivotal role in steering the construction industry toward a sustainable development trajectory.

The sustainable potential bearing capacity evaluation of tourism environments is a critical multiple-attribute group decision-making (MAGDM) problem, particularly in the context of achieving "dual carbon" goals. This chapter explores the application of the evaluation based on distance from average solution (EDAS) technique under intuitionistic fuzzy sets (IFSs) to address uncertainty in decision-making. By integrating Hamming and logarithmic distance measures, the proposed IFN-EDAS method enhances the assessment of tourism environments, considering factors such as ecological, economic, and industrial development potential. The criteria importance through intercriteria correlation (CRITIC) technique is employed to derive objective attribute weights, ensuring a robust evaluation framework [1]. A case study demonstrates the method's effectiveness in prioritizing sustainable tourism destinations, aligning with low-carbon development objectives. Comparative analyses with existing techniques (e.g., induced fuzzy weighted averaging (IFWA), induced fuzzy weighted geometric (IFWG), ViseKriterijumska Optimizacija I Kompromisno Resenje (VIKOR)) validate the model's reliability, highlighting its utility in balancing environmental sustainability with tourism growth. This approach contributes to the MADM literature by offering a systematic tool for policymakers to optimize resource allocation and promote eco-friendly tourism practices.

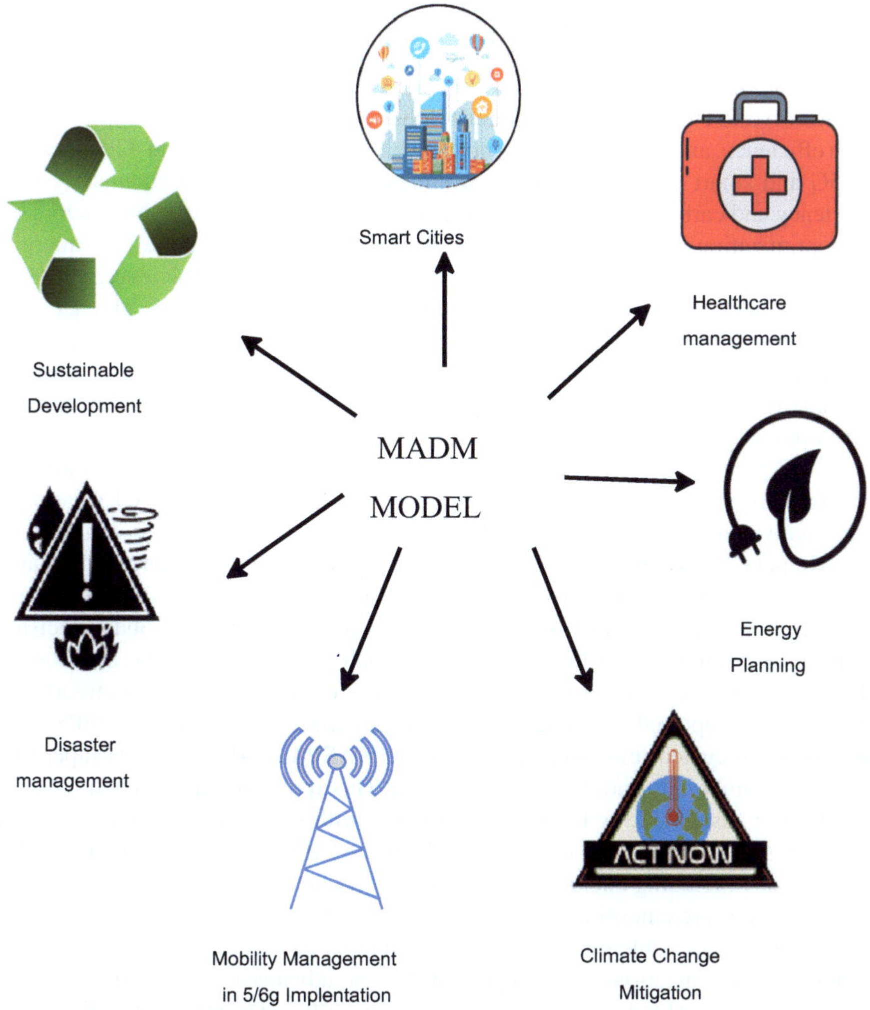

Fig. 8.1 Application of MADM

8.2 Smart Cities

The rapid advancement of 5G networks, coupled with the rise of smart cities and the Internet of Things (IoT), has underscored the need for sustainable urban development. As outlined in the Brundtland Report, sustainability requires meeting present needs without compromising future generations—a principle that is now being integrated into the design of next-generation wireless networks. 5G technology, with its enhanced mobile broadband (eMBB), massive machine-type communications (mMTC), and ultra-reliable low-latency communication (uRLLC), serves as a critical enabler for smart city applications such as intelligent transportation, smart grids, and remote healthcare. However, the increased demand for high-speed connectivity and massive IoT deployments also raises challenges in energy consumption, spectrum efficiency, and environmental impact [3]. Research indicates that 42% of studies on 5G and smart cities focus on environmental sustainability, particularly energy efficiency and carbon footprint reduction, while 37% examine economic factors like cost optimization and 21% address social aspects such as health and safety. Emerging technologies like network slicing, edge computing, and renewable energy integration are helping mitigate these challenges, paving the way for smarter, greener urban ecosystems. Moving forward, the convergence of 5G with future 6G networks and non-terrestrial communication systems will further enhance sustainability, ensuring that smart cities not only improve quality of life but also align with global sustainable development goals.

MADM methods play a crucial role in addressing complex urban challenges by evaluating multiple criteria to optimize smart city initiatives. A key application involves assessing Internet of Things (IoT) adoption barriers in smart waste management (SCWM), where Fermatean fuzzy sets (FFSs) enhance uncertainty modeling. The FF-combined compromise solution (CoCoSo) method, integrated with Archimedean copula-based aggregation operators, enables robust decision-making by balancing conflicting objectives such as cost, efficiency, and sustainability. Case studies demonstrate its effectiveness in prioritizing smart city alternatives, ensuring optimal resource allocation. Comparative analyses with TOPSIS and weighted aggregated sum product assessment (WASPAS) [4] validate its superiority in handling ambiguity and dynamic urban data. This approach supports policymakers in selecting sustainable IoT solutions, aligning with smart city development goals in Fig. 8.2. Future research may explore hybrid MADM models for broader urban applications, including energy, mobility, and governance.

Smart cities are emerging as a transformative solution to address global environmental challenges while enhancing citizens' quality of life. By leveraging Information and Communication Technologies (ICTs) and advanced decision support tools, municipalities worldwide are optimizing energy efficiency and sustainability [5]. These systems enable real-time monitoring of key performance indicators (KPIs) and data-driven decision-making to improve urban governance. A notable example is the Urban Control Centre (UCC) in Bari, Italy, which integrates multiple energy-related subsystems such as public lighting, residential buildings, and transportation into a unified decision-making platform. Unlike traditional sector-specific approaches, the UCC adopts a holistic strategy, coordinating conflicting objectives across urban sectors to maximize efficiency and sustainability. This innovative framework demonstrates how smart cities can harmonize environmental goals with civic needs through integrated, ICT-enabled governance.

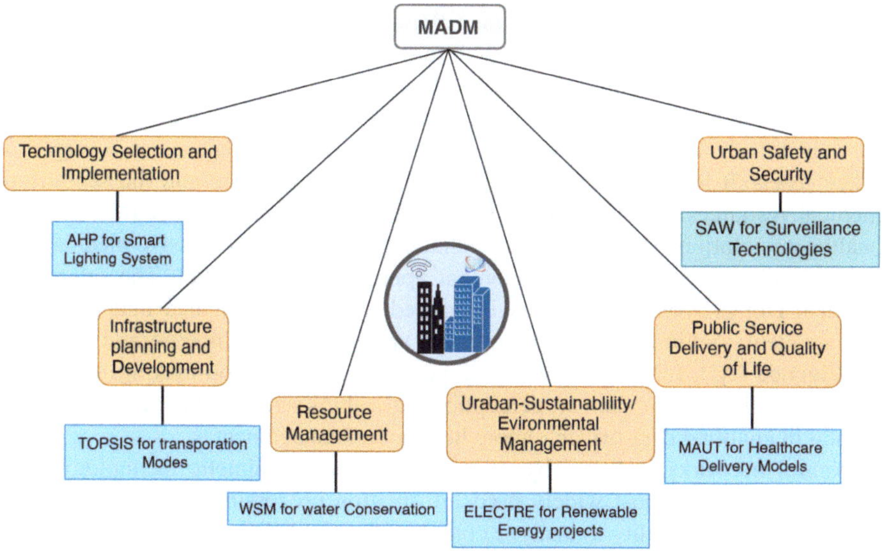

Fig. 8.2 Smart city

8.3 Healthcare Management

Healthcare waste management is a critical challenge that demands sustainable solutions to mitigate risks to public health and the environment. Medical facilities generate hazardous waste including infectious materials, sharps, pharmaceuticals, and chemicals, which require specialized disposal methods to prevent disease transmission and ecological damage. Traditional approaches like incineration, while effective for pathogen destruction, often produce toxic emissions, while landfilling risks groundwater contamination. To address these complex decisions, researchers have developed advanced multi-criteria decision-making models like the intuitionistic hesitant fuzzy MULTIMOORA (IHF-MULTIMOORA) method, which incorporates uncertainty in expert evaluations through membership and non-membership degrees. This approach evaluates disposal alternatives across three dimensions—ratio analysis, reference point comparison, and full multiplicative form—to determine optimal solutions for different waste types: deep burial for pathological waste, incineration for pharmaceuticals, autoclaving for sharps, and chemical disinfection for hazardous chemicals [6]. The growing adoption of sustainable technologies like microwave treatment and steam sterilization, coupled with improved segregation practices using color-coded bins, demonstrates progress in healthcare waste management. However, developing nations still face significant challenges including inadequate infrastructure, limited funding, and weak regulatory enforcement. Future solutions will likely integrate emerging technologies like AI-powered waste tracking and 6G-enabled remote monitoring systems, while policy initiatives must focus on capacity building, stricter compliance with WHO guidelines, and circular economy approaches to reduce waste generation. Effective healthcare waste management ultimately requires balancing technological innovation, environmental protection, and public health priorities through evidence-based decision frameworks.

With the advancement of human society, public concerns have shifted from basic survival to physical healthcare. In recent years, smart medicine has emerged as a key research focus in academia. The COVID-19 pandemic and the scarcity of medical resources in remote areas have further heightened the need for efficient healthcare solutions. To address these challenges, leveraging massive medical data through robust, data-driven management approaches is essential. The Internet of Medical Things (IoMT) has enabled integrated platforms for medical data fusion and management, yielding numerous successful applications. By incorporating artificial intelligence (AI), IoMT systems facilitate intelligent decision-making in medical data processing. Consequently, perception algorithms within IoMT frameworks are pivotal in enabling data-driven remote healthcare management.

Neural networks, inspired by neuroscience and biomimetic technology, have become one of the most widely used information processing architectures in the past decade. Their ability to analyze complex patterns has made them invaluable in medical healthcare management, with significant progress demonstrated in various studies. Some works even integrate decision algorithms with IoMT technologies to better align with real-world applications. However, most existing approaches rely on single-view models, which are often tailored to specific scenarios and lack adaptability in sub-optimal medical data conditions.

To overcome these limitations, this work introduces a multi view deep learning framework for medical data analytics. By combining different neural network models from multiple perspectives, we enhance feature representation and improve prediction accuracy in healthcare forecasting [7]. Our approach fuses three distinct neural network models to create a comprehensive mapping for final predictions, effectively addressing the shortcomings of traditional single-model systems. This method increases the value density of raw medical data, making it more suitable for large-scale healthcare management tasks.

Some people propose the Multiview Deep Learning-Based Medical Data Management (MDL-MDM) framework, focusing on survival time prediction for cancer patients a key regression problem in medical data analysis. The framework consists of two main components: an IoMT platform architecture for secure data collection and transmission, and a multi view neural network workflow for intelligent healthcare predictions. The MDL-MDM system integrates data access through wearable sensors and medical devices, wireless data transmission via protocols like Wi-Fi and 5G, advanced data analysis using multi view neural networks, and hospital side data integration for decision support. This holistic approach ensures accurate, adaptable, and secure healthcare management. Key contributions of this work include identifying challenges in data-driven medical management for remote monitoring, proposing a novel multi view deep learning framework for survival time prediction, and validating its performance through extensive experiments. The proposed framework demonstrates significant potential in enhancing medical data utilization and improving patient outcomes in smart healthcare systems. The emerging Internet of Things (IoT) technology offers transformative solutions for healthcare through real-time remote monitoring systems. This chapter presents a fault-tolerant mHealth framework (FTF-mHealth-IoT) designed to address critical failures in telemedicine architecture, particularly in wearable sensors (Tier 1) and medical Centre servers (Tier 3). The framework incorporates a risk-level localization triage (RLLT) algorithm to autonomously detect patient emergencies and identify risk-level healthcare service packages, eliminating dependency on centralized triage control. Additionally, it employs MCDM specifically the AHP to evaluate and prioritize hospitals based on service availability and time-to-arrival (TAH). Validated using datasets from 572 chronic heart disease patients and 24 hospitals across Baghdad and Kuala Lumpur, the framework demonstrates robust performance in optimizing hospital selection under varying service availability scenarios. Key contributions include enhanced fault tolerance, real-time decision-making, and improved healthcare delivery efficiency, supporting both patients and medical staff in modern healthcare ecosystems. Future directions include extending the framework to multiple chronic diseases and implementing real-time adaptive decision platforms.

The rapid advancement of Internet of Things (IoT) technologies has revolutionized healthcare by enabling real-time remote monitoring and telemedicine solutions, yet critical challenges such as wearable sensor failures and medical server downtimes can disrupt patient care, particularly for high-risk individuals. To address these vulnerabilities, this chapter presents a fault-tolerant mhealth framework (FTF-mHealth-IoT) that ensures uninterrupted healthcare services through intelligent decision-making and redundancy mechanisms. The framework incorporates a risk-level localization triage (RLLT) algorithm that autonomously detects emergencies by analyzing real-time patient data and self-reported symptoms while providing early failure alerts for malfunctioning sensors. For optimal care coordination, it employs multi criteria decision making (MCDM), specifically the AHP, to evaluate and prioritize hospitals based on service availability and time-to-arrival (TAH), dynamically bypassing centralized servers during failures. Validated with 572 chronic heart disease patients and 24 hospitals across different service availability scenarios, the framework demonstrates high reliability in triage accuracy, efficient hospital prioritization, and scalability for diverse healthcare infrastructures. By integrating wearable sensors, mHealth apps, and cloud-based systems, it enables bidirectional communication for seamless data flow. The implications for healthcare management include enhanced patient-centric care through timely interventions, improved operational resilience by reducing central system dependency, and cost-efficient resource allocation. Future directions involve extending the framework to multi-disease monitoring, implementing adaptive decision platforms for dynamic patient conditions, and incorporating emergency first-aid guidance [8]. Ultimately, FTF-mHealth-IoT represents a significant advancement in IoT-driven healthcare, combining fault tolerance, real-time analytics, and intelligent decision-making to support sustainable, efficient, and patient focused telemedicine systems.

8.4 Mobility Management in 5G Implementation

Mobility management in 5G networks represents a critical challenge as user equipment moves across increasingly complex heterogeneous environments. The handover process, particularly vertical handovers between different radio access technologies, requires sophisticated decision-making frameworks to maintain quality of service while minimizing disruptions. Traditional signal-strength-based approaches have proven inadequate for 5G's diverse use cases, prompting the adoption of MADM methods that evaluate multiple quality indicators simultaneously. These advanced techniques, including TOPSIS, VIKOR, and GRA, analyze up to ten network parameters from latency and bandwidth to energy consumption and cost through a three-stage process of normalization, weighting, and ranking. The integration of fuzzy logic and machine learning with MADM methods has further enhanced their ability to handle real-world uncertainties and predict optimal handover timing. However, challenges remain in reducing computational overhead for ultra-dense networks and standardizing these approaches across vendor implementations. As networks evolve toward 6G, mobility management systems must address emerging requirements like AI-driven predictive handovers and energy-efficient algorithms for massive IoT deployments while maintaining the seamless connectivity expected by users of bandwidth-intensive applications like augmented reality and autonomous vehicles [9]. The future of mobility management lies in adaptive frameworks that can balance technical constraints with quality of experience across increasingly diverse network environments. The deployment of 5G wireless networks brings unprecedented improvements in speed, latency, and connectivity, but also introduces complex challenges in mobility management especially within heterogeneous networks (HetNets) that integrate macro cells and a dense overlay of small cells. These architectures, while enhancing network coverage and capacity, result in frequent handovers, increased handover failures, ping-pong effects, and elevated signaling overhead. Traditional handover mechanisms based on single-criterion decisions, such as received signal strength alone, fall short in such dynamic environments. This is where MCDM techniques play a crucial role.

MCDM approaches such as the AHP, GRA, and TOPSIS enable more intelligent and holistic handover decisions by simultaneously considering multiple performance indicators. These include parameters like received signal strength indicator (RSSI), signal-to-interference-plus-noise ratio (SINR), cell load, user speed, and Quality of Service (QoS) requirements. By assigning appropriate weights to each metric, MCDM techniques help select the most suitable target cell, reducing unnecessary handovers and enhancing mobility robustness. One prominent application is in beam management for millimeter-wave (mmWave) systems, were rapid signal fluctuations due to obstacles and high user mobility demand frequent beam realignments. MCDM can be employed to evaluate candidate beams based on stability and quality, thereby improving communication reliability. Additionally, load balancing is enhanced as MCDM-based schemes can dynamically adjust handover thresholds and cell individual offsets (CIO) to distribute traffic evenly across network tiers. This not only improves resource utilization but also contributes to energy-efficient operation. Emerging approaches combine MCDM with software-defined networking (SDN) and dual connectivity (DC) to further optimize mobility management. SDN facilitates centralized control and global network awareness, allowing MCDM algorithms to make real-time, context-aware decisions. DC ensures connection reliability by maintaining simultaneous links to macro and small cells during mobility events [10]. Future research directions point toward integrating MCDM with machine learning (ML) models for adaptive and predictive handover decisions, and employing conditional handover (CHO) techniques to proactively mitigate millimeter-wave communication challenges. MCDM offers a robust framework for addressing the multifaceted challenges of mobility management in 5G networks. Its ability to incorporate diverse criteria into the decision-making process ensures improved handover performance, enhanced user experience, and greater network efficiency making it a critical enabler of seamless connectivity in next-generation wireless systems.

8.5 Energy Planning

The planning stage of any project, including energy supply systems, is described as having crucial significance and involves contributions from field experts to decision-makers (DM). The specific model presented in the sources is for planning of energy sources for a microgrid using a MCDM approach based on the AHP [11]. This planning process is undertaken to choose the best alternative from available options. The model was developed considering the electrical energy load demand of an educational institution as a reference. The planning process considers the main-utility grid as the primary source, alongside conventional sources like diesel generator (DG) and gas-based combined heat-and-power (CHP) with absorption chiller, as well as renewable energy sources like solar photovoltaic (PV) together with battery energy storage system (BESS). The assessment and evaluation for prioritization of these energy sources is based on critical criteria or attributes and their associated sub-criteria. The critical influencing criteria focused on in this model include economic, technical, structural, operational and maintenance, environmental, and social aspects.

In this planning model, nine alternative combinations of the grid and other energy source (s) are identified to form the microgrid. The objective is the forming of the microgrid. The process involves three identifying alternatives, defining criteria, computing weights incorporating relevant criteria and sub criteria, and finally evaluating alternatives to find the best suited one. Figure 8.3 presents all the phases of the MCDM process employed in formulating energy planning through microgrid policy. It outlines each step involved, demonstrating how multi-criteria decision-making supports the evaluation and selection of optimal strategies within the microgrid policy framework. A practical application of this framework is demonstrated through a case study of an academic institution in India with 125 kW peak demand and annual consumption of 375,000 kWh. The analysis reveals that a grid-connected photovoltaic system (score: 0.136) emerges as the optimal solution, offering the best balance between sustainability and scalability. Close alternatives include hybrid systems combining grid, PV, battery storage, and combined heat/power (CHP) units (score: 0.130), which provide enhanced reliability through islanding capability during grid outages. The framework highlights several critical insights for energy planners [11]. First, it quantifies the trade-off between higher initial investments in renewable technologies and their long-term economic and environmental benefits. Second, it demonstrates how microgrid solutions can address both current energy needs and future expansion requirements through modular design. Third, it provides a structured approach to incorporate often-overlooked factors like social acceptance and local skill availability into technical decisions. Looking ahead, the energy planning process must evolve to incorporate dynamic elements. Future enhancements to the MCDM framework could integrate real-time data on energy prices, weather patterns, and demand fluctuations. The growing importance of policy factors like carbon pricing and renewable incentives suggests the need for scenario-based analysis capabilities. Emerging technologies like AI could further refine the decision process by analyzing historical performance data to optimize weight assignments.

The transition to sustainable energy systems represents one of the most significant infrastructure challenges of our time. By adopting structured decision-making frameworks like the MCDM approach presented here, organizations can navigate this complexity with greater confidence [11]. The methodology provides not just a tool for selecting energy solutions today, but a flexible platform for continuous evaluation as technologies, costs, and priorities evolve. For energy planners, policymakers, and facility managers, such frameworks offer a pathway to develop energy systems that are simultaneously economically viable, technically robust, environmentally sustainable, and socially responsible the essential requirements for meeting both current needs and future challenges.

Energy planning presents complex multi-dimensional challenges that require balancing competing objectives such as cost efficiency, system reliability, and environmental sustainability. MCDM has emerged as a powerful analytical framework to address these challenges, as demonstrated through two representative case studies. In the first case involving generation expansion planning for a U.S. utility, the AHP was employed to systematically evaluate alternatives across three policy dimensions: SO_2 emissions control strategies, demand-side management adoption, and system reliability levels. The analysis revealed that while the CDL strategy (combining demand-side management, emissions control, and low unserved energy) emerged as optimal under base assumptions, sensitivity analysis identified competitive alternatives when accounting for real-world uncertainties in both data measurements (±5%) and priority weights (±10%). The second case study applied MCDM to design a hybrid renewable energy system, where AHP helped identify robust configurations that balanced cost (50.1% weight), reliability (30.9%), and emissions (19.0%) across multiple future scenarios [12]. These applications demonstrate MCDM's unique value in energy planning: its structured approach enables transparent evaluation of trade-offs, systematic incorporation of stakeholder preferences through criteria weighting, and rigorous assessment of solution robustness under uncertainty. The methodology's flexibility across different energy contexts—from traditional utility

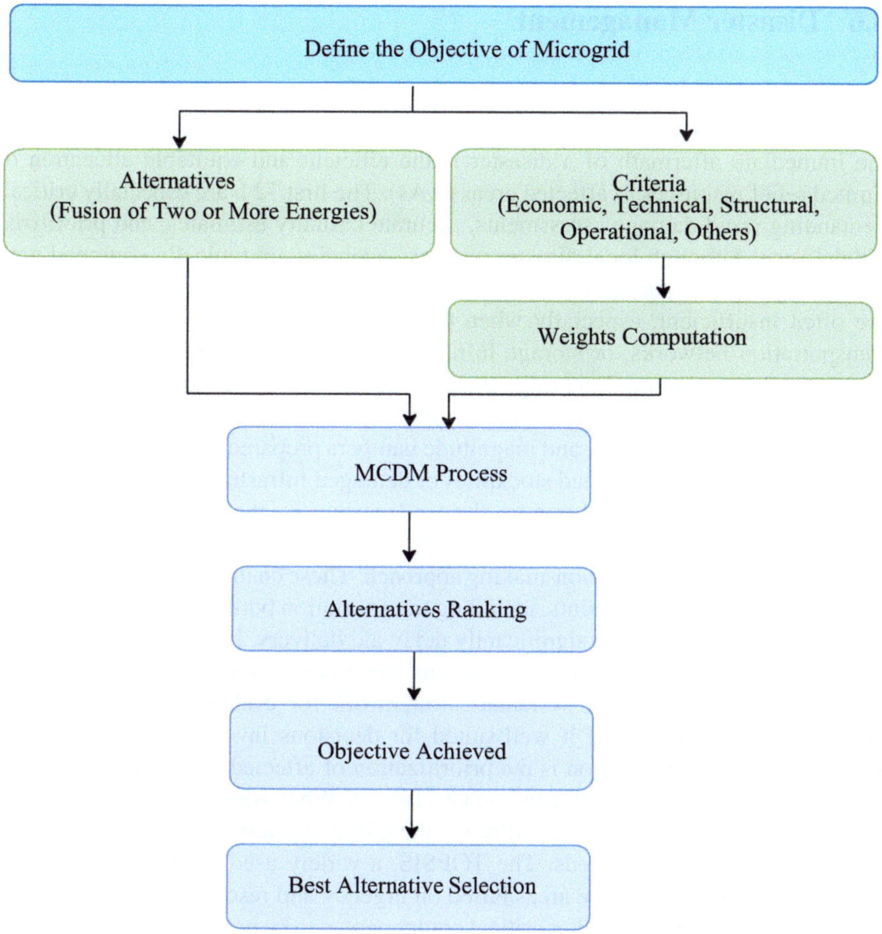

Fig. 8.3 Energy planning using MCDM

planning to renewable energy integration—combined with its ability to handle both quantitative and qualitative factors, makes it particularly valuable for addressing the complex, multi-stakeholder decisions characteristic of modern energy systems. Future applications could benefit from integrating MCDM with emerging techniques like machine learning to enhance scenario analysis and uncertainty quantification in energy transition planning.

8.6 Disaster Management

Disasters, by their very nature, occur with little to no warning, resulting in widespread damage, disruption, and human suffering. One of the most pressing challenges in the immediate aftermath of a disaster is the efficient and equitable allocation of limited relief resources to affected areas (AAs). The first 72 h are especially critical, demanding rapid damage assessments, accurate casualty estimates, and prioritized aid delivery. Although local disaster response agencies are typically equipped with emergency supplies such as food, medical kits, and rescue equipment, their resources are often insufficient, especially when the disaster impairs communication lines, transportation networks, or storage infrastructure. In such scenarios, international support becomes essential to bridge the gap between available and required aid.

Several challenges compound the complexity of disaster response. The unpredictability of disaster timing and magnitude hampers preparedness efforts. Resource shortages, either due to limited stockpiles or damaged infrastructure, can leave local agencies overwhelmed. Furthermore, the need to prioritize the most severely affected areas such as regions with high injury rates or critical infrastructure damage calls for a structured and rational decision-making approach. These challenges are further exacerbated by logistical constraints, including transportation bottlenecks and communication breakdowns that can significantly delay aid delivery. In this context, MCDM techniques play a pivotal role in improving the effectiveness of disaster response strategies. MCDM offers a systematic framework for evaluating multiple, often conflicting, criteria making it well-suited for decisions involving complex trade-offs. One primary application is the prioritization of affected areas. Here, information crowdsourced from platforms such as Twitter, WhatsApp, and Ushahidi can be harnessed to assess real-time conditions, including casualty figures, infrastructure damage, and immediate needs. The TOPSIS, a widely used MCDM method, can then be applied to rank these areas based on urgency and resource requirements.

Resource allocation, another critical component, can be optimized using the solid transportation problem (STP), an extension of classical transportation models that accounts for capacity constraints and priority-based delivery. This allocation operates within a three-echelon supply chain structure. In the first echelon, local relief distribution centers (RDCs) address the most urgent needs. In the second echelon, central disaster coordination centers (CDCs) fulfill the residual demand. Finally, the third echelon ensures that RDCs are replenished for sustained operations. This layered approach not only ensures the timely delivery of aid but also supports long-term recovery. Cost optimization is another key consideration in humanitarian logistics. Relief operations must strive for a balance between efficiency and effectiveness minimizing operational costs while maximizing lives saved. Integrated models aim to reduce transportation, processing, and deprivation costs, while still ensuring equitable and timely aid distribution across all affected regions.

A computational case study focusing on earthquake response in Tripura, India, demonstrates the practical application of these concepts. The simulation involved local RDCs addressing high-priority zones delivering, for instance, 80% of medical supplies to areas with the most injuries. CDCs covered the remaining needs and restocked RDCs as necessary. Crowdsourced data, processed through TOPSIS, guided the ranking and prioritization of AAs, resulting in an optimized and effective relief effort. Managerial insights drawn from such models reveal the critical role of real-time demand estimation through crowdsourcing, the benefits of priority-based distribution strategies, and the cost advantages of optimized routing and inventory management [13]. These practices enhance situational awareness, improve survival outcomes, and reduce the financial burden of disaster response, enabling agencies to allocate resources toward long-term recovery efforts.

Looking ahead, several advancements can further refine these approaches. Incorporating fuzzy MCDM models can better handle uncertainties in demand and supply estimation. Integrating two-phase evacuation strategies that consider both medical urgency and transportation logistics could improve rescue efficiency. Enhancing route resilience models to adapt to damaged infrastructure and deploying hybrid algorithms powered by artificial intelligence can further increase the speed and accuracy of decision-making. The integration of MADM methods into disaster management provides a robust, data-driven framework for making informed decisions in high-stakes scenarios. By combining real-time data, optimization techniques, and structured evaluation models, MCDM enhances the speed, fairness, and effectiveness of humanitarian response ultimately saving lives and strengthening resilience against future disasters.

Disaster management involves complex decision-making under uncertainty, requiring the evaluation of multiple, often conflicting criteria such as timeliness, resource allocation, and risk mitigation. The MCDM methodologies provide a structured framework to address these challenges by systematically assessing alternatives based on weighted criteria [14]. In disaster response, MCDM aids in prioritizing emergency actions, optimizing resource distribution, and coordinating multi-agency efforts. For instance, techniques like the AHP or TOPSIS can evaluate evacuation routes, relief supply chains, or communication strategies by integrating quantitative and qualitative factors, such as infrastructure resilience, population density, and response time. By leveraging MCDM, stakeholders can enhance situational awareness, improve coordination, and ensure efficient resource utilization, ultimately minimizing human and economic losses. The integration of MCDM with emerging technologies, such as UAV-assisted communication networks, further strengthens disaster resilience by enabling data-driven, adaptive decision-making in dynamic environments. Thus, MCDM serves as a critical tool in transforming fragmented disaster management into a holistic, evidence-based process.

MCDM approaches play a crucial role in disaster management by providing structured frameworks to evaluate complex, uncertain scenarios with multiple conflicting criteria. In typhoon disaster assessment, where data may be incomplete or ambiguous, advanced techniques like trapezoidal fuzzy neutrosophic numbers (TrFNNs) effectively capture truth, indeterminacy, and falsity degrees in damage evaluations, outperforming traditional fuzzy methods. The integration of GRA further enhances decision-making by correlating disaster impact factors while preserving nuanced information, avoiding premature data simplification [15]. A case study on typhoon assessment in Fujian, China, demonstrated how this approach objectively weights criteria using entropy methods and realistically handles missing data without misleading assumptions. By combining TrFNNs with GRA, this MCDM framework enables more accurate impact assessments and prioritization of relief efforts, ultimately improving disaster response effectiveness. The methodology's adaptability suggests promising applications for other natural disasters, provided it incorporates evolving data sources like real-time AI analytics for comprehensive situational awareness.

8.7 Climate Change Mitigation

Climate change mitigation has garnered significant research attention, with MCDM methods playing a pivotal role in evaluating and prioritizing sustainable strategies. Recent studies have applied MCDM techniques to assess emission reduction policies, renewable energy adoption, and sector-specific mitigation efforts, such as cement production, agriculture, and urban transportation [16]. For instance, researchers have utilized advanced frameworks like type-2 neutrosophic sets to handle uncertainty in decision-making, integrating methods such as MEREC for objective criteria weighting and MARCOS for alternative ranking. Comparative analyses with TOPSIS and WASPAS confirm the reliability of these approaches, demonstrating their effectiveness in real-world applications. By systematically evaluating trade-offs between environmental, economic, and social criteria, MCDM provides policymakers with robust tools to implement sustainable climate actions, ensuring balanced and evidence-based mitigation strategies. This highlights the critical role of structured decision-making frameworks in addressing the complexities of climate change. Mitigating climate change requires strategic interventions across diverse sectors, balancing environmental goals with economic and social considerations. MCDM methods offer structured approaches to evaluate these complex trade-offs, guiding the development of effective mitigation strategies [17]. Within the context of energy systems, MCDM supports the integration of renewable sources like wind, solar, and hydropower into power grids, ensuring stability and efficiency through distributed control systems and dynamic pricing models. In the transportation sector, optimization-based MCDM helps prioritize low-emission alternatives such as electric vehicles, hydrogen fuels, and smart mobility solutions by weighing emissions reduction, energy consumption, and infrastructure readiness. For industrial decarbonization, MCDM facilitates the deployment of Industry 4.0 technologies including digital twins and AI-driven optimization to reduce emissions in energy-intensive processes like steel and cement production.

MCDM also plays a pivotal role in evaluating negative emissions technologies, such as carbon capture, utilization, and storage (CCUS), by assessing their economic viability, environmental impact, and implementation risks. These evaluations help policymakers align mitigation strategies with national and global net-zero commitments. The incorporation of emerging approaches such as controls co-design, human-in-the-loop frameworks, and stakeholder-centric models ensures that MCDM systems remain adaptive, inclusive, and responsive to real-world challenges. By enabling systematic assessment of competing objectives, MCDM empowers decision-makers to develop robust, technically sound, and socially equitable climate mitigation strategies paving the way for a low-carbon future.

8.8 Summary

The provided content highlights the critical role of MCDM in addressing complex challenges across 5G/6G mobility management, energy planning, disaster management, and climate change mitigation. In 5G/6G networks, MCDM techniques such AHP, GRA, and TOPSIS optimize handover decisions by evaluating multiple factors such as signal strength, interference, and network load, ensuring seamless connectivity in ultra-dense and heterogeneous environments. Future advancements include AI-driven predictive handovers and SDN integration for adaptive mobility management. In energy planning, MCDM frameworks balance economic, technical, and environmental factors, with case studies demonstrating the effectiveness of hybrid systems like grid-connected solar PV. For disaster management, MCDM aids in rapid resource allocation and prioritization of affected areas using real-time data and optimization models, improving emergency response efficiency. Lastly, in climate change mitigation, MCDM evaluates renewable energy adoption, emission reduction strategies, and industrial decarbonization, leveraging advanced techniques like neuromorphic sets and AI for robust decision-making. Overall, MCDM provides a systematic, data-driven approach to navigating trade-offs and enhancing sustainability, resilience, and efficiency across these critical domains.

References

1. R. Yan, Y. Han, F. Li, P. Li, Evaluation of sustainable potential bearing capacity of tourism environment under uncertainty: a multiphase intuitionistic fuzzy EDAS technique based on hamming distance and logarithmic distance measures. IEEE Access **12**, 8081–8095 (2024)
2. Q. Li, Y. Li, A combined MAGDM-based framework employing EDAS and critic techniques for green building technology schemes evaluation. IEEE Access **11**, 135120–135132 (2023)
3. M.J. Shehab, I. Kassem, A.A. Kutty, M. Kucukvar, N. Onat, T. Khattab, 5g networks towards smart and sustainable cities: a review of recent developments, applications and future perspectives. IEEE Access **10**, 2987–3006 (2021)
4. A.R. Mishra, P. Rani, A. Saha, I.M. Hezam, D. Pamucar, M. Marinović, K. Pandey, Assessing the adaptation of internet of things (IoT) barriers for smart cities' waste management using fermatean fuzzy combined compromise solution approach. IEEE Access **10**, 37109–37130 (2022)

5. R. Carli, M. Dotoli, R. Pellegrino, ICT and optimization for the energy management of smart cities: The street lighting decision panel, in *2015 IEEE 20th Conference on Emerging Technologies Factory Automation (ETFA)* (2015), pp. 1–6
6. S. Geetha, S. Narayanamoorthy, D. Kang, J.V. Kureethara, A novel assessment of healthcare waste disposal methods: intuitionistic hesitant fuzzy multimoora decision making approach. IEEE Access **7**, 130283–130299 (2019)
7. K. Yu, L. Quan, C. Chakraborty, X. Qi, Y. Shen, Z. Guo, O. Alfarraj, A. Tolba, Multiview deep learningbased efficient medical data management for survival time forecasting. IEEE J. Biomed. Health Inform. pp. 1–13 (2024)
8. O.S. Albahri, A.S. Albahri, A.A. Zaidan, B.B. Zaidan, M.A. Alsalem, A.H. Mohsin, K.I. Mohammed, A.H. Alamoodi, S. Nidhal, O. Enaizan, et al., Fault-tolerant mhealth framework in the context of iot-based real-time wearable health data sensors. IEEE Access **7**, 50052–50080 (2019)
9. F.S. Dantas Silva, M.P.S. Lima, D. Corujo, A.J. Venâncio Neto, F. Esposito, A comprehensive step-wise survey of multiple attribute decision-making mobility approaches. IEEE Access **12**, 108616–108656 (2024)
10. E. Gures, I. Shayea, A. Alhammadi, M. Ergen, H. Mohamad, A comprehensive survey on mobility management in 5g heterogeneous networks: Architectures, challenges and solutions. IEEE Access **8**, 195883–195913 (2020)
11. S.S. Bohra, A. Anvari-Moghaddam, B. Mohammadi-Ivatloo, Ahp-assisted multi-criteria decision-making model for planning of microgrids, in *IECON 2019—45th Annual Conference of the IEEE Industrial Electronics Society*, vol. 1 (2019), pp. 4557–4562 (2019)
12. J. Pan, Y. Teklu, S. Rahman, A. de Castro, An interval-based MADM approach to the identification of candidate alternatives in strategic resource planning. IEEE Trans. Power Syst. **15**(4), 1441–1446 (2000)
13. D. Sarma, A. Das, P. Dutta, U.K. Bera, A cost minimization resource allocation model for disaster relief operations with an information crowdsourcing-based MCDM approach. IEEE Trans. Eng. Manag. **69**(5), 2454–2474 (2022)
14. G. Fragkos, E.E. Tsiropoulou, S. Papavassiliou, Disaster management and information transmission decision-making in public safety systems, in *2019 IEEE Global Communications Conference (GLOBECOM)* (2019), pp. 1–6
15. R. Tan, W. Zhang, S. Chen, Decision-making method based on grey relation analysis and trapezoidal fuzzy neutrosophic numbers under double incomplete information and its application in typhoon disaster assessment. IEEE Access **8**, 3606–3628 (2019)
16. V. Simic, I. Gokasar, M. Deveci, L. Švadlenka, Mitigating climate change effects of urban transportation using a type-2 neutrosophic merec-marcos model. IEEE Trans. Eng. Manag. **71**, 3233–3249 (2024)
17. P.P. Khargonekar, T. Samad, S. Amin, A. Chakrabortty, F. Dabbene, A. Das, M. Fujita, M. Garcia-Sanz, D.F Gayme, M.D. Ilić, et al., Climate change mitigation, adaptation, and resilience: challenges and opportunities for the control systems community. IEEE Control. Syst. Mag. **44**(3), 33–51 (2024)

MIX
Papier aus verantwortungsvollen Quellen
Paper from responsible sources
FSC® C105338

If you have any concerns about our products,
you can contact us on
ProductSafety@springernature.com

In case Publisher is established outside the EU,
the EU authorized representative is:
Springer Nature Customer Service Center GmbH
Europaplatz 3, 69115 Heidelberg, Germany

Printed by Libri Plureos GmbH
in Hamburg, Germany